# Lecture Notes in Computer Science 4977

Commenced Publication in 1973
Founding and Former Series Editors:
Gerhard Goos, Juris Hartmanis, and Jan van Leeuwen

Yoshiharu Ishikawa   Jing He   Guandong Xu
Yong Shi   Guangyan Huang   Chaoyi Pang
Qing Zhang   Guoren Wang (Eds.)

# Advanced Web and Network Technologies, and Applications

APWeb 2008 International Workshops:
BIDM, IWHDM, and DeWeb
Shenyang, China, April 26-28, 2008
Revised Selected Papers

 Springer

Volume Editors

Yoshiharu Ishikawa
Nagoya University, Nagoya, Japan
E-mail: ishikawa@itc.nagoya-u.ac.jp

Jing He
Yong Shi
CAS Research Center on Data Technology and Knowledge Economy, Beijing, China
E-mail: {hejing, yshi}@gucas.ac.cn

Guandong Xu
Victoria University, Melbourne, Australia
E-mail: xu@csm.vu.edu.au

Guangyan Huang
Institute of Software, Chinese Academy of Sciences, Beijing, China
E-mail: huanggy@ercist.iscas.ac.cn

Chaoyi Pang
Qing Zhang
CSIRO ICT Centre, Brisbane, QLD, Australia
E-mail: {chaoyi.pang, qing.zhang}@csiro.au

Guoren Wang
Northeastern University, Shenyang, China
E-mail: wanggr@mail.neu.edu.cn

Library of Congress Control Number: 2008938735

CR Subject Classification (1998): H.3, H.4, H.5, C.2, K.4

LNCS Sublibrary: SL 3 – Information Systems and Application, incl. Internet/Web
and HCI

ISSN       0302-9743
ISBN-10    3-540-89375-X Springer Berlin Heidelberg New York
ISBN-13    978-3-540-89375-2 Springer Berlin Heidelberg New York

Springer is a part of Springer Science+Business Media

springer.com

© Springer-Verlag Berlin Heidelberg 2008
Printed in Germany

Typesetting: Camera-ready by author, data conversion by Scientific Publishing Services, Chennai, India
Printed on acid-free paper    SPIN: 12568604    06/3180    5 4 3 2 1 0

# APWeb 2008 Workshop Chairs' Message

It is our pleasure to present the proceedings of the 10th Asia Pacific Web Conference Workshops held at Shenyang, China during April 26-28, 2008. The APWeb workshop is an international forum for Web and information technology researchers and practitioners to exchange their ideas regarding innovations in the state of art and practice of Web and information technology, as well as to identify the promising research topics and characterize the future of information technology research. This year, four workshops were held in conjunction with APWeb 2008, namely, Information Explosion and Next-Generation Search, The First Workshop on Business Intelligence and Data Mining, International Workshop on Health Data Management 2008, and Doctoral Consortium On Data Engineering and Web Technology Research. The APWeb workshop has proven to be an excellent catalyst for further research and collaboration, and we fully expect that this year's meeting continued this trend. This year's workshop program featured a variety of papers, focusing on topics ranging from Web searching, Web services, database, data mining, bioinformatics, and business intelligence. These topics play very important roles in creating next-generation information technology architectures and solutions.

The workshop attracted submission from seven countries. The submissions received and also contained in this volume reflect the international character of APWeb workshops. We would like to thank all workshop Chairs for providing their service to the community; it has been an honor to work with the workshop Program Committee for their thoughtful and erudite reviews of the papers. The foremost thanks, however, is due to all the authors who submitted their work for the workshop; the workshop Program Committee faced a difficult task in selecting the papers ultimately accepted for presentation.

APWeb 2008 supported the above four workshops to help enhance the communications between researchers and funding organizations; an industry panel provided guidance for future development foci and interests. Another core objective was the initiation of spirited discourse and, hopefully, of collaborations among those at the workshop.

June 2008

Yoshiharu Ishikawa
Jing He

# Business Intelligence and Data Mining Workshop Chairs' Message

Intelligent data analysis provides powerful and effective tools for problem solving in a variety of business modeling tasks. The first workshop on business intelligence and data mining (BIDM 2008) focused on data science aspects of banking and financial risk management, social networks that relate to business intelligence. It included but was not limited to modeling, numeric computation, algorithmic and complexity issues in data mining technology for risk management, credit assessment, social network, asset/derivatives interest rate determination, insurance, foreign exchange rate forecasting, online auction, cooperative game theory, general equilibrium, information pricing, network bandwidth pricing, rational expectation, repeated games, etc. Paper submissions by both tool developers and users from the scientific and engineering community were encouraged in order to inspire communication between both groups.

Organized in conjunction with the 10th Asia Pacific Web Conference (APWeb 2008), BIDM 2008 aimed to publish and disseminate knowledge on an international basis in the areas of business intelligence, intelligent data analysis, and data mining. It provided a forum for state-of-the-art developments and research, as well as current innovative activities in business intelligence, data analysis, and mining.

BIDM attracted submission from four countries. The four research papers contained in this volume were selected by the Program Committee (PC) in a double-blind review process from a total of ten submissions. These papers cover a wide range of both theoretical and pragmatic issues related to the application of data analysis and mining techniques in business. The submissions reflect the international character of the workshop.

The six papers in this volume include four research papers and two keynote papers. Of the two keynote papers, one is by Zhiming Ding from the Software Institute of the Chinese Academy of Sciences and the other by Xijin Tang from the Academy of Mathematics and System Sciences, Chinese Academy of Sciences.

The workshop was partially sponsored by the National Nature Science Foundation of China (No. 70602034, 70531040, 70472074). It was our pleasure to serve as Co-chairs for this workshop. We would like to take this opportunity to express our appreciations to all PC members for their expertise and hard work in evaluating papers.

June 2008

Yong Shi
Guangyan Huang
Jing He

# Health Data Management Workshop Chairs' Message

Information collected during the treatment of disease is an under-utilized resource in medical and health research. The management of this collected data for the purpose of health research is a fundamental question which is only now attracting greater interest. This management needs to start at the point of collection and requires the gathering of patient consent through the integration of the data from multiple information systems, fathering of information from related data sets, to the analysis of the combined data. If done properly, this process has the ability to lead to improvements to the quality of patient care, prevention of medical errors, and reduction of healthcare costs. There is much to be achieved with the realization of this goal, both in terms of theoretical information management research and the development of practical and useful applications. The collaborations of database researchers and clinical experts are important for this research.

Organized in conjunction with the 10th Asia Pacific Web Conference (APweb 2008), the purpose of IWHDM 2008 was to provide a forum for discussion and interaction among researchers with interests in the cutting-edge issues of information technology techniques in health. IWHDM 2008 attracted submissions from five countries and each paper was carefully reviewed by at least three members of an international Program Committee (PC). The six papers in this volume include four research papers and two keynote papers: one keynote paper by Gary Morgan from the Australian E-Health Research Centre and the other by Yucai Feng from the DaMeng Database Company. These papers cover a wide range of both theoretical and pragmatic issues related to IT in Healthcare.

The workshop was sponsored by the Australian E-Health Research Centre and APWEB 2008. We were pleased to serve as PC chairs for this workshop and would like to take this opportunity to express our appreciations to all PC members for their expertise and help in evaluating papers.

June 2008

Chaoyi Pang
Qing Zhang

# Doctoral Consortium on Data Engineering and Web Technology Research Workshop Chairs' Message

The Doctoral Consortium on Data Engineering and Web Technology Research (DeWeb 2008) is a premium younger researcher forum, which aims to bring together current PhD students working on the topics related to the APWeb conference. DeWeb 2008 provided an opportunity for doctoral students to present their research topics and early results and receive advice from research experts in these areas. The consortium operated in a workshop format to provide guidance for each student's future progress. The consortium was one of the four workshops running in conjunction with the APWeb 2008 conference.

This year's DeWeb workshop attracted 22 research papers and each paper was reviewed by at least two international Program Committee members or invited reviewers. Based on the technical comments from peer reviewers on the aspects of originality, significance, technical quality, relevance, clarity of presentation, seven quality papers were selected to be presented at the workshop. Meanwhile, four invited papers which reported their latest research progresses on the related topics made by four distinguished researchers are also included in this proceedings volume.

We would like to take this opportunity to thank all the authors who submitted their papers to the workshop. We also thank the Program Committee members and numerous invited reviewers for their volunteer work on paper review. We are also grateful to the conference organizers for their help.

June 2008

Guoren Wang
Guandong Xu

# Organization

## Organization Committee

Yong Shi — Chinese Academy of Sciences, China
Guangyan Huang — Chinese Academy of Sciences, China
Jing He — Chinese Academy of Sciences, China
Chaoyi Pang — CSIRO ICT Centre, Australia
Qing Zhang — CSIRO ICT Centre, Australia
Guoren Wang — Northeastern University, China
Guandong Xu — Victoria University, Australia

## Program Committee

### The First Workshop on Business Intelligence and Data Mining

Tianshu Wang — China Research Lab, IBM
Aihua Li — Central University of Finance and Economics
Ying Bao — Industrial and Commercial Bank of China

### International Workshop on Health Data Management

Chaochang Chiu — YuanZe University, Taiwan
Frédérique Laforest — Lyon Research Centre for Images and Information Systems, France
Weiyi Liu — Yunnan University, China
Stefan Jablonski — University of Bayreuth, Germany
Ilias Maglogiannis — University of the Aegean, Greece
Natalia Sidorova — Technical University Eindhoven, The Netherlands
Yuval Shahar — Ben-Gurion University of the Negev, Israel
Xuequn Shang — Northwestern Polytechnic University, China
Bing Wu — Dublin Institute of Technology, Ireland
Kun Yue — Yunnan University, China

### Doctoral Consortium on Data Engineering and Web Technology Research

Chengfei Liu — Swinburne University of Technology, Australia
Lei Chen — Hong Kong University of Science and Technology, Hong Kong

# Table of Contents

## The First Workshop on Business Intelligence and Data Mining

Moving Objects Databases Based on Dynamic Transportation
Networks: Modeling, Indexing, and Implementation .................. 1
    *Zhiming Ding*

Approach to Detection of Community's Consensus and Interest ........ 17
    *Xijin Tang*

A Comparative Empirical Study on the Margin Setting of Stock Index
Futures Calendar Spread Trading ................................. 30
    *Haizhen Yang, Hongliang Yan, and Ni Peng*

A Study on Multi-word Extraction from Chinese Documents .......... 42
    *Wen Zhang, Taketoshi Yoshida, and Xijin Tang*

Extracting Information from Semi-structured Web Documents: A
Framework ...................................................... 54
    *Nasrullah Memon, Abdul Rasool Qureshi, David Hicks, and
    Nicholas Harkiolakis*

Discovering Interesting Classification Rules with Particle Swarm
Algorithm....................................................... 65
    *Yi Jiang, Ling Wang, and Li Chen*

## International Workshop on Health Data Management

Improving the Use, Analysis and Integration of Patient Health Data.... 74
    *David Hansen, Mohan Karunanithi, Michael Lawley,
    Anthony Maeder, Simon McBride, Gary Morgan, Chaoyi Pang,
    Olivier Salvado, and Antti Sarela*

DM-Based Medical Solution and Application ....................... 85
    *Junjie Liang and Yucai Feng*

Learning-Function-Augmented Inferences of Causalities Implied in
Health Data..................................................... 87
    *JiaDong Zhang, Kun Yue, and WeiYi Liu*

Support Vector Machine for Outlier Detection in Breast Cancer
Survivability Prediction.......................................... 99
    *Jaree Thongkam, Guandong Xu, Yanchun Zhang, and Fuchun Huang*

An Empirical Study of Combined Classifiers for Knowledge Discovery
on Medical Data Bases ........................................... 110
  *Lucelene Lopes, Edson Emilio Scalabrin, and Paulo Fernandes*

Tracing the Application of Clinical Guidelines ....................... 122
  *Eladio Domínguez, Beatriz Pérez, and María A. Zapata*

## Doctoral Consortium on Data Engineering and Web Technology Research

The Research on the Algorithms of Keyword Search in Relational
Database ........................................................ 134
  *Peng Li, Qing Zhu, and Shan Wang*

An Approach to Monitor Scenario-Based Temporal Properties in Web
Service Compositions ............................................. 144
  *Pengcheng Zhang, Bixin Li, Henry Muccini, and Mingjie Sun*

Efficient Authentication and Authorization Infrastructure for Mobile
Users ........................................................... 155
  *Zhen Ai Jin, Sang-Eon Lee, and Kee Young Yoo*

An Effective Feature Selection Method Using the Contribution
Likelihood Ratio of Attributes for Classification..................... 165
  *Zhiwang Zhang, Yong Shi, Guangxia Gao, and Yaohui Chai*

Unsupervised Text Learning Based on Context Mixture Model with
Dirichlet Prior .................................................. 172
  *Dongling Chen, Daling Wang, and Ge Yu*

The Knowledge Discovery Research on User's Mobility of
Communication Service Provider ................................... 182
  *Lingling Zhang, Jun Li, Guangli Nie, and Haiming Fu*

Protecting Information Sharing in Distributed Collaborative
Environment ..................................................... 192
  *Min Li and Hua Wang*

Relevance Feedback Learning for Web Image Retrieval Using Soft
Support Vector Machine ........................................... 201
  *Yifei Zhang, Daling Wang, and Ge Yu*

Feature Matrix Extraction and Classification of XML Pages ........... 210
  *Hongcan Yan, Dianchuan Jin, Lihong Li, Baoxiang Liu, and
  Yanan Hao*

Tuning the Cardinality of Skyline .................................... 220
  *Jianmei Huang, Dabin Ding, Guoren Wang, and Junchang Xin*

An HMM Approach to Anonymity Analysis of Continuous Mixes ...... 232
  *Zhen Ling, Junzhou Luo, and Ming Yang*

**Author Index** ................................................. 245

# Moving Objects Databases Based on Dynamic Transportation Networks: Modeling, Indexing, and Implementation

Zhiming Ding

Institute of Software, Chinese Academy of Sciences,
South-Fourth-Street 4, Zhong-Guan-Cun, Beijing 100080, P.R. China
zhiming@iscas.ac.cn

**Abstract.** In this paper, a new moving objects database model, Route-based model for Moving Objects on Dynamic Transportation Networks (RMODTN), is proposed which is suited for dealing with the interrelationship between moving objects and the underlying transportation networks. The data model is given as a collection of data types and operations which can be plugged into a DBMS to obtain a complete data model and query language. Besides, an index framework for network-constrained moving objects is provided, which can deal with the full dynamic trajectories of network-constrained moving objects so that the queries on the whole life span of the moving objects can be efficiently supported. To evaluate the performance of the proposed methods, we have implemented RMODTN on PostgreSQL and conducted a series of experiments. The experimental results show quite satisfying query processing performances.

**Keywords:** Database, Spatio-temporal, Moving Objects, Algebra, Index.

## 1 Introduction

Moving Objects Database (MOD) is the database that can track and manage the continuously changing locations of moving objects such as cars, ships, flights, and pedestrians. Combined with dynamically updated location information and other information such as spatial data and location dependent data, MOD can answer many important and interesting queries, such as "*tell me the nearest k taxi cabs around my current position*", "*which places did Tom visit yesterday afternoon*".

In recent years, the management of moving objects has been intensely investigated. In [12, 16], Wolfson *et al.* have proposed a Moving Objects Spatio-Temporal (MOST) model which is capable of tracking not only the current, but also the near future positions of moving objects. Su *et al.* in [14] have presented a data model for moving objects based on linear constraint databases. In [9], Güting *et al.* have presented a data model and data structures for moving objects based on abstract data types. Besides, Pfoser and Jensen *et al.* in [11] have discussed the indexing problem for moving object trajectories. However, nearly none of the above work has treated the interaction between moving objects and the underlying traffic networks in any way.

Y. Ishikawa et al. (Eds.): APWeb 2008 Workshops, LNCS 4977, pp. 1–16, 2008.
© Springer-Verlag Berlin Heidelberg 2008

More recently, increasing research interests are focused on modeling transportation networks and network constrained moving objects. Papadias *et al.* in [10] have presented a framework to support spatial network databases. Vazirgiannis *et al.* in [15] have discussed moving objects on fixed road networks. In [13], the authors have presented a computational data model for network-constrained moving objects. Besides, the index problems of network constrained moving objects have also been studied [1, 5]. In [6], the authors have proposed a fixed-network based moving objects database model with a rich set of data types and operations defined. However, all these works have only considered static transportation networks, so that topology changes and state changes of the network can not be expressed.

Moreover, in [6] the transportation network is defined as a single data type with an interface to relations and standard data types provided, which may affect the efficiency in dealing with dynamic situations. Besides, the model is focused on abstract data types and utilizes the sliced representation for the discrete model without considering the relationship between location updates and moving object trajectories. This can cause problems since a new slice is needed whenever the moving object changes its speed or direction so that huge data can be yielded.

In [2], the authors have proposed an edge-based dynamic transportation network model, which can present state and topology changes. The main problem with the model is that edge-based network frameworks are not suitable for location update purposes, since a new location update is needed whenever the moving object runs to a different edge, which can cause a big location update overhead. In [3], the authors have proposed a dynamic network based moving object model. However, the model is not suited for the database framework since the data types and operations are not defined. Besides, the location update strategies are based on mile-meters instead of GPS so that it can have a lot of limitations in real-world applications.

To solve the above problems, we propose a new moving objects database model, Route-based model for Moving Objects on Dynamic Transportation Networks (RMODTN), in this paper. In RMODTN, the traffic network framework is route-based, which differs from the edge-based model defined in [2] and is suitable for location update purposes. Besides, data types and operations are defined directly on detailed network components (such as routes and junctions), different from the network model in [6] which defines the whole network as a single data type.

The remaining part of this paper is organized as follows. Section 2 formally defines the RMODTN model, including data types and operations; Section 3 propose an index structure for network-constrained moving objects, Section 4 discuss implementation issues and performance evaluation results, and Section 5 finally concludes the paper.

## 2   RMODTN Database Model

In this section, we will formally define the RMODTN model. Obviously, to make this model readable and clean, it is crucial to have a formal specification framework which allows us to describe widely varying data models and query languages. Such a specification framework, called second-order-signature, was proposed in [7]. The basic idea is to use a system of two coupled signatures where the first signature describes a type system and the second one describes an algebra over the types of the first signature. In the following discussion, we will define our model with the second-order-signature. Especially, we will focus on discrete model so that the data types and operations defined in this paper can be implemented directly in an extensible database system such as PostgreSQL or Secondo [8]. The notation of the definitions will follow those described in [9].

The methodology proposed in this paper arose from the observation that in most real-life applications, moving objects only move inside existing transportation networks instead of moving arbitrarily in the X×Y×Z space. This means that we can actually model moving objects on the predefined transportation networks instead of modeling them directly in the Euclidean space.

Besides, in the whole MOD system, there can be a lot of general events which are hard to be expressed by the states of moving objects alone, such as traffic jams, blockages caused by temporary constructions, and changes to the topology of the transportation networks. Therefore, we model the underlying transportation networks as dynamic graphs which allow us to express these events. For simplicity, "dynamic transportation networks" and "dynamic graphs" will be used interchangeably throughout this paper.

## 2.1 Data Types

Table 1 presents the type system of the RMODTN model. Type constructors listed in Group 1 are basic ones which have been defined and implemented in [9]. In the following discussion, we mainly focus on the type constructors listed in Group 2.

**Table 1.** Signatures describing the type system of RMODTN

| Group | Type constructor | Signature | |
|---|---|---|---|
| | _int_, _real_, _string_, _bool_ | | → BASE |
| | _point_, _points_, _line_, _region_ | | → SPATIAL |
| 1 | _instant_ | | → TIME |
| | _range_ | BASE ∪ TIME | → RANGE |
| | _intime_, _moving_ | BASE ∪ SPATIAL | → TEMPORAL |
| | _blockage_, _blockreason_, _blockpos_ | | → GBLOCK |
| | _statedetail_, _state_ | | → GSTATE |
| | _temporalunit_, _temporal_, _intimestate_ | | → GTEMPORAL |
| 2 | _dynroute_, _dyninjunct_, _dyninterjunct_, _dyngaph_ | | → GRAPH |
| | _gpoint_, _gpoints_, _grsect_, _gline_, _gregion_ | | → GSPATIAL |
| extending → _intime_, _moving_ | | {_gpoint_}∪BASE∪SPATIAL | → TEMPORAL |

As shown in Table 1, temporal data types are obtained by "extending" the previously defined type constructors _moving_ and _intime_ to include _gpoint_ as their arguments. SPATIAL data types are still reserved for these two type constructors to deal with the situations when moving objects move outside of the predefined transportation networks.

### 2.1.1  Graph State Data Types and Graph Blockage Data Types

**Definition 1 (state).** The carrier set of the *state* data type is defined as follows:

$$D_{state} = \{\text{opened, closed, blocked}\}$$

**Definition 2 (blockage reason).** The data type *blockreason* describes the reason of a blockage. Its carrier set is defined as follows:

$$D_{blockreason} = \{\text{temporal-construction, traffic-jam, car-accident, undefined}\}$$

**Definition 3 (interval).** Let $(S, <)$ be a set with a total order. Intervals and closed intervals over $S$ can be defined as follows:

$$interval(S) = \{(s, e, lc, rc) \mid s,\, e \in S,\, lc,\, rc \in \text{bool},\, s \le e,\, (s=e) \Rightarrow (lc=rc=true)\}$$
$$cinterval(S) = \{(s, e, lc, rc) \mid s,\, e \in S,\, s \le e,\, lc=rc=true\}$$

where *lc* and *rc* are two flags indicating "left-closed" and "right-closed" respectively.

**Definition 4 (blockage position).** The data type *blockpos* is used to describe the position of a blockage, and its carrier set is defined as follows:

$$D_{blockpos} = \{\, \Psi \mid \Psi \in cinterval([0,1])\}$$

In Definition 4, we use a closed interval over [0, 1] to indicate the blockage position, whose boundaries indicate the border of the blocked area. Suppose that the total length of the route is 1, and then any location in the route can be represented by a real number $p \in [0, 1]$.

**Definition 5 (blockage, blockages).** The *blockage* data type is used to describe a blockage, including its reason and its location. The *blockages* data type describes multiple blockages inside one single route. Their carrier sets are defined as follows:

$$D_{blockage} = \{(br, \Psi) \mid br \in D_{blockreason},\, \Psi \in D_{blockpos} \}$$

$$D_{blockages} = \{B \mid B \subseteq D_{blockage} \}$$

**Definition 6 (state detail).** The data type *statedetail* is used to describe the detailed state of a junction or a route, and its carrier set is defined as follows:

$$D_{statedetail} = \{(s, B) \mid s \in D_{state},\, B \in D_{blockages},\, s \ne \text{blocked} \Leftrightarrow B = \varnothing\}$$

### 2.1.2  Graph Temporal Data Types

Graph temporal data types are used to track the state history and also the life span of a junction or a route.

**Definition 7 (temporal unit).** The *temporalunit* data type describes the state of a junction or a route during a certain time period. Its carrier set is defined as follows:

$$D_{temporalunit} = \{(I, sd) \mid I \in interval(D_{instant}),\, sd \in D_{statedetail} \}$$

**Definition 8 (temporal).** The *temporal* data type is defined as a sequence of temporal units which describe the state history of a junction or a route:

$$D_{temporal} = \{< \mu_1, \ldots, \mu_n > \mid n \ge 1,\ \mu_i = (I_i, sd_i) \in D_{temporalunit}\ (1 \le i \le n),\ \text{and:}$$

(1) $\forall i, j \in \{1, \dots n\}, i \neq j: I_i \cap I_j = \varnothing$

(2) $\forall i \in \{1, \dots n\text{-}1\}: I_i \lhd I_{i+1}$ ($\lhd$ means "before" in time series)}

The insertion and deletion time of a junction or a route can be decided by $\min(I_1)$ and $\max(I_n)$ respectively. In this way, we can track the topology of the graph system.

**Definition 9 (intimestate).** The *intimestate* data type is used to describe the state of a junction or a route at a certain time instant. Its carrier set is defined as follows:

$$D_{\underline{intimestate}} = \{(t, sd) \mid t \in D_{\underline{instant}}, sd \in D_{\underline{statedetail}} \}$$

### 2.1.3 Dynamic Graph Data Types

In RMODTN, transportation networks are modeled as dynamic graphs, with every junction or route associated with a *temporal* attribute which describes its state history.

**Definition 10 (dynamic route).** A dynamic route can be viewed as a normal graph route with a temporal attribute associated. The carrier set of the *dynroute* data type is:

$$D_{\underline{dynroute}} = \{( rid, geometry, len, tp) \mid rid \in D_{\underline{int}}, route \in polyline, len \in D_{\underline{real}}, tp \in D_{\underline{temporal}}\}$$

where *rid* is the identifier of the route which is isomorphic to integer, *len* is the length of the route, *tp* is the temporal attribute associated with the route, and *geometry* is a polyline which describes the geographical shape of the route. In this way, a dynamic route can actually assume a shape of complicated curve in the X×Y plane instead of just a straight line. The polyline is considered as directed, whose direction is from the first vertex to the last vertex, which enables us to speak of the beginning point (or 0-end) and the end point (or 1-end) of the route.

**Definition 11 (dynamic in-graph junction).** A dynamic in-graph junction can be considered as normal junction with a temporal attribute associated, which connects two or more routes of the same graph. The carrier set of the *dyninjunct* data type is:

$$D_{\underline{dyninjunct}} = \{ (jid, loc, ((rid_i, pos_i))_{i=1}^{n}, m, tp) \mid jid \in D_{\underline{int}}, loc \in D_{\underline{point}}, tp \in D_{\underline{temporal}} \}$$

where *jid* is the identifier of the dynamic junction, *loc* is a point value which describes the position of the junction, *m* is the connectivity metrix [3] which describes the connectivity of the junction, and *tp* is the temporal attribute associated with the junction. $(rid_i, pos_i)$ $(1 \leq i \leq n)$ in the above definition indicates the *i*th route connected by the junction, where $rid_i$ is the identifier of the route and $pos_i \in [0, 1]$ describes the position of the junction inside the route.

**Definition 12 (dynamic graph).** A dynamic graph, $G$, is composed of a set of dynamic routes and a set of dynamic in-graph junctions. The carrier set of the *dyngraph* data type is defined as follows:

$$D_{\underline{dyngraph}} = \{(gid, R, J) \mid gid \in D_{\underline{int}} , R \subseteq D_{\underline{dynroute}}, J \subseteq D_{\underline{dyninjunct}} \}$$

In implementation, $R$ and $J$ can be implemented as relational tables so that in a *dyngraph* value only the relation names are contained.

**Definition 13 (dynamic inter-graph junction).** A dynamic inter-graph junction is a junction which connects routes from different graphs. The carrier set of the *dyninterjunct* data type is defined as follows:

$$D_{dyninterjunct} = \{(jid, loc, ((gid_i, rid_i, pos_i))_{i=1}^{n}, m, tp) \mid jid \in D_{int}, loc \in D_{point}, tp \in D_{temporal}\}$$

The definition of the inter-graph junction is very similar to that of the in-graph junction. The 3-tuple $(gid_i, rid_i, pos_i)$ $(1 \leq i \leq n)$ describes the routes connected by the inter-graph junction, which can come from different graphs.

### 2.1.4 Graph Spatial Data Types

Based on the above definitions for dynamic transportation networks, we can then define some useful data types, *graph point*, *graph points*, *graph route section*, *graph line*, and *graph region*, which form the basis for the modeling and querying of moving objects.

**Definition 14 (graph point, graph points).** The *gpoint* data type describes a point inside the graph system, and the *gpoints* data type describes a set of graph points:

$$D_{gpoint} = \{( gid, rid, pos) \mid gid, rid \in D_{int}, pos \in [0, 1]\}$$

$$D_{gpoints} = \{PS \mid PS \subseteq D_{gpoint}\}$$

**Definition 15 (graph route section).** The *grsect* data type represents a section of a route. Its carrier set is defined as follows:

$$D_{grsect} = \{(gid, rid, S) \mid gid, rid \in D_{int}, S \in cinterval([0, 1])\}$$

**Definition 16 (graph line).** A graph line is defined as a consecutive chain of route sections inside the graph system. Its carrier set is defined as follows:

$$D_{gline} = \{<\omega_i>_{i=1}^{n} \mid n \geq 1, \omega_i = (gid_i, rid_i, S_i) \in D_{grsect}, and:$$

$$(1)\forall i \in \{1, n\}: S_i \in cinterval([0, 1]); (2) \ \forall i \in \{1, ...n-1\} : adjacent(\omega_i, \omega_{i+1})\}$$

where $adjacent(\omega_i, \omega_{i+1})$ means that $\omega_i, \omega_{i+1}$ meet with their end points spatially so that all route sections of the graph line can form a chain.

**Definition 17 (graph region).** A graph region is defined as an arbitrary set of graph route sections. The carrier set of the *gregion* data type is defined as follows:

$$D_{gregion} = \{ W \mid W \subseteq D_{grsect}\}$$

### 2.1.5 Temporal Data Types

In [9], Güting *et al.* have defined the *moving* and *intime* type constructors which take BASE data types and SPATIAL data types as arguments. In the following we extend these two type constructors by taking the *gpoint* data type also as its argument and define the *moving*(*gpoint*) and *intime*(*gpoint*) data types.

Conceptually, a moving graph point *mgp* can be defined as a function from time to graph point: $mgp = f: D_{instant} \rightarrow D_{gpoint}$. In implementation, we should translate the above definition into a discrete representation. That is, a moving graph point is expressed as a series of motion vectors, and each motion vector describes the movement of the moving object at a certain time instant.

**Definition 18 (motion vector).** The carrier set of the data type *mvector* is:

$$D_{mvector} = \{(t, gp, \vec{v}) \mid t \in D_{instant}, gp=(gid, rid, pos) \in D_{gpoint}, \vec{v} \in D_{real}\}$$

where $\vec{v}$ is the speed measure. Its absolute value is equal to the speed of the moving object, while its sign (either positive or negative) depends on the direction of the moving object. If the moving object is moving from 0-end towards 1-end, then the sign is positive. Otherwise, if it is moving from 1-end to 0-end, the sign is negative.

**Definition 19 (moving graph point).** A moving graph point can be represented by a sequence of motion vectors. The carrier set of the *moving(gpoint)* data type (or *mgpoint* for short) can be defined as follows:

$$D_{mgpoints} = \{ (\delta_i)_{i=1}^{n} \mid n \geq 1, \delta_i = (t_i, gp_i, \vec{v}_i) \in D_{mvector} \ (1 \leq i \leq n), \text{ and:}$$

$$\forall i \in \{1, \dots n-1\}: t_i \lhd t_{i+1} \}$$

For a running moving object, its motion vectors are generated by location updates (including IDTLU, DTTLU, and STTLU, see [3, 4]), and the last motion vector, called "active motion vector", contains the current moving pattern of the moving object, which is the key information in computing the current (or near future) location of the moving object and in triggering the next location update.

Through the sequence of motion vectors, the location of the moving object at any time instant during its life span can be computed from its motion vectors through interpolation. Therefore, a moving graph point value can be viewed as a spatial-temporal trajectory which is composed of a sequence of trajectory units. For two consecutive motion vectors $mv_s$ and $mv_e$, the trajectory unit is denoted as $\mu(mv_s, mv_e)$, which is a line segment linking $mv_s$ and $mv_e$ in the spatial-temporal space. For the active motion vector $mv_a$, the corresponding trajectory unit, denoted as $\mu(mv_a)$, is called "active trajectory unit", which is a radial starting from $mv_a$.

**Definition 20.** The *intime(gpoint)* data type is used to describe a graph position of a moving object at a certain time instant, and its carrier set is defined as follows:

$$D_{intime(gpoint)} = \{(t, gp) \mid t \in D_{instant}, gp \in D_{gpoint}\}$$

## 2.2 Operations

In the earlier work on moving objects databases [9], Güing *et al.* have defined and implemented a rich set of operations on the data types listed in Group 1 of Table 1. In this subsection, we will show how these predefined operations can be systematically adapted to the RMODTN model by an "Extending" technique. Table 2 gives a summary of the operations in the RMODTN model.

In designing RMODTN operations, the general rules for "extending" can be summarized as follows: (1) Every operation whose signature involves *point* is extended to include *gpoint* also; (2) Every operation whose signature involves *line* is extended to include *grsect* and *gline* also; (3) Some of the operations whose signature involves

**Table 2.** Operations of the RMODTN Model

| Group | Class | Operations |
|-------|-------|------------|
| Non-Temporal | Predicates | isempty, $\equiv$ , $\not\equiv$ , $<$ , $\leq$ , $>$ , $\geq$ , intersects, inside, before touches, attached, overlaps, on_border, in_interior |
| | Set operations | intersection, union, minus, crossings, touch_points, common_border |
| | Aggregation | min, max, avg, avg[center], single |
| | Numeric | no_components, size, perimeter, size[duration], size[length], size[area] |
| | Distance & direction | distance, direction |
| | Base type specific | and, or, not |
| Temporal | Projection to Domain/Range | deftime, rangevalues, locations, trajectory, routes, traversed, inst, val |
| | Interaction with Domain/Range | atinstant, atperiods, initial, final, present, at, atmin, atmax, passes |
| | When | when |
| | Lifting | (All new operations inferred) |
| | Rate of Change | derivative, speed, turn, velocity |
| Graph Specific | Transformation | graph_euc, euc_graph, dyngraph, dyninjunction, dyninterjunction, dynroute, getid |
| | Construction | gpoint, grsect |
| | Data Extraction | getjunctions, getroutes, pos, route, temporal, atinstant, statedetail, state, blockages, blocksel, blockpos |
| | Truncation | atperiods, present, at |
| | When | when |
| | Projection | deftime |

*region* are extended to include *gregion* also; and (4) Operations which are only suited for 1D data types (see [9]) or other specific data types (such as *region*) are not extended.

According to the above rules, the underscored (line-underscored and dot-underscored) operations in Table 2 are extended while other operations are not affected.

First let's deal with the non-temporal operations listed in Table 2. In [9], the signatures of most non-temporal operations (see the line-underscored non-temporal operations, such as **isempty**) are defined with two data type variables $\pi$ and $\sigma$, where $\pi \in \{int, bool, string, real, instant, point\}$ and $\sigma \in \{range(int), range(bool), range(string), range(real), periods, points, line, region\}$. Now we extend the domain of $\pi$ and $\sigma$ like this:

$\pi \in \{int, bool, string, real, instant, point\} \cup \{gpoint\}$

$\sigma \in \{range(int), range(bool), range(string), range(real), periods, points, line, region\} \cup \{gpoints, grsect, gline, gregion\}$.

As a result of this change, all operations whose signatures are defined with $\pi$, $\sigma$ variables are extended to cover the newly introduced data types automatically.

As for the temporal operations listed in Table 2, the extension can be made in a similar way. In [9], the signatures of most temporal operations (see the line-underscored temporal operations, such as **<u>deftime</u>**) are defined by two data type variables $\alpha$ and $\beta$, ($\alpha$, $\beta \in$ BASE$\cup$SPATIAL). Now we extend the domain of $\alpha$ and $\beta$ like this:

$\alpha \in$ BASE$\cup$SPATIAL$\cup\{$<u>*gpoint*</u>$\}$

$\beta \in$ BASE$\cup$SPATIAL$\cup$GSPATIAL

By this extension, the operations whose signatures are defined with $\alpha$, $\beta$ are extended automatically.

As for operations whose signatures are not defined with $\pi$, $\sigma$, $\alpha$, $\beta$ variables (see the dot-underscored operations in Table 2, such as **crossings** and **locations**), we have to add some supplementary signatures. The basic rule for adding supplementary signatures is that <u>*gline*</u> and <u>*gregion*</u> are semantically equivalent to <u>*line*</u>, while <u>*gpoint*</u> is semantically equivalent to <u>*point*</u>.

Through the above extension, the previously defined operations are enabled to deal with the data types newly introduced in this paper. The above extension is also suited to the Lifted operations.

In addition to the extended operations, we also define a set of new operations, which are mainly focused on graph specific data types (see Table 2). The signatures of graph specific operations take the same forms as described in [2].

### 2.3 Query Examples

Based on the data types defined above, we can then define database schemas with the new data types as attributes. For instance, we can have the following schemas:

Hagenroutes (name: *string*, droute: *dynroute*);
Hagenjunctions (name: *string*, djunction: *dyninjunct*);
Movingobjs (mname: *string*, mid: *int*, mgp: *mgpoint*);

We can also use the operations defined in this paper in the SQL statements (suppose MILE, METER, and MINUTE are constant real number values which represent the values of one mile, one meter, and one minute respectively).

**Example 1.** "Find all cargos that are currently within 5 miles from my position p".

Select mname, mid
From   movingobjs
Where  **distance**(**val**(**atinst**(mgp, NOW)), p)<=5 * MILE;

**Example 2.** "Find all cargo pairs which are within 100m with each other for more than 5 minutes"

Select A.mid, B.mid
From   movingobjs A, movingobjs B
Where  **duration**(**at**(**distance**(A.mgp, B.mgp),[0, 100*METER]))>5*MINUTE;

This query shows the join of moving objects. The result of **distance**(A.mgp, B.mgp) is a moving real value.

## 3   Indexing the Full Trajectories of Network Constrained Moving Objects

In this section, we describe the index method for network-constrained moving objects in RMODTN. The structure of the index, called *Network-constrained moving objects Dynamic Trajectory R-Tree* (NDTR-Tree), is two-layered. The upper layer is a single R-Tree and the lower layer consists of a forest of R-Trees, similar to MON-Tree [1].

However, different from [1, 5], NDTR-Tree employs a hybrid structure. Its upper R-Tree is edge-based, that is, the basic unit for indexing is the edges with smaller granularity, so that the intersection between different MBRs can be greatly reduced; while its lower R-Trees are route-based with each lower R-Tree corresponding to a route which has a greater granularity, so that location update and index maintaining costs can be reduced. In this way, the query processing and index maintaining performances can be improved. Since each route can contain multiple edges with each edge connecting two neighboring junctions, multiple leaf records of the upper R-Tree can point to the same lower R-Tree, as shown in Figure 1.

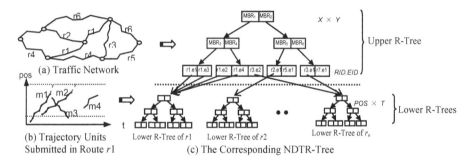

**Fig. 1.** Structure of the NDTR-Tree

Besides, the NDTR-Tree can index the full trajectories of moving objects, and by maintaining the index dynamically when location updates occur, the NDTR-Tree can keep the latest location information in the index structure so that queries on historical, present, and near future locations of moving objects can be supported.

In the following, let's first consider the structure of NDTR-Tree. We suppose that the function route(*rid*) returns the route corresponding to the specified identifier, and the function RTree$_{low}$(*rid*) returns the corresponding lower R-Tree of route(*rid*).

In the upper R-Tree, the records of the leaf nodes take the form <*MBR$_{xy}$*, *rid.eid*, *pt$_{route}$*, *pt$_{tree}$*>, where *MBR$_{xy}$* is the *Minimum Bounding Rectangle* (MBR) (in the *X×Y* plane) of the edge, *rid.eid* is a combination of the route and edge identifiers which uniquely specifies an edge of a route, *pt$_{route}$* is a pointer to the detailed route record, and *pt$_{tree}$* is a pointer to the lower R-Tree corresponding to route(*rid*). The root or internal nodes contain records of the form <*MBR$_{xy}$*, *pt$_{node}$*>, where *MBR$_{xy}$* is the MBR (in the *X×Y* plane) containing all MBRs of its child records, and *pt$_{node}$* is a pointer to the child node.

The lower part of the NDTR-Tree is composed of a set of R-Trees with each R-Tree corresponding to a certain route and indexing all the trajectory units submitted by the moving objects in the route. The root or internal nodes of the lower R-Tree contains records of the form $<MBR_{pt}, pt_{node}>$, where $MBR_{pt}$ is the MBR (in the $POS \times T$ plane) containing all MBRs of its child records, and $pt_{node}$ is the pointer to its child node. The records of the leaf nodes have the form $<MBR_{pt}, mid, mv_s, mv_e>$, where $MBR_{pt}$ is the MBR (in the $POS \times T$ plane) bounding $\mu(mv_s, mv_e)$, $mid$ is the identifier of the moving object, $mv_s = (t_s, rid, pos_s, \vec{v}_s)$ and $mv_e = (t_e, rid, pos_e, \vec{v}_e)$ are the two consecutive motion vectors which form the trajectory unit (if $\mu(mv_s, mv_e)$ is active trajectory unit, then $mv_e$ is null).

In deciding the MBR of a trajectory unit $\mu(mv_s, mv_e)$, if $\mu(mv_s, mv_e)$ is a non-active trajectory unit, then the MBR is $<t_s, pos_s, t_e, pos_e>$. If $\mu(mv_s, mv_e)$ is an active trajectory unit, then we only need to predict to the end of the route with the moving object running with the slowest speed $(|\vec{v}_s - \psi|)$ ($\psi$ is the speed threhold).Therefore, its MBR is $<t_s, pos_s, t_{end}, pos_{end}>$, where $t_{end}$ and $pos_{end}$ can be computed as follows (we assume the moving object runs towards 1-end):

$$t_{end} = t_s + \frac{(1 - pos_s) * r.length}{|\vec{v}_s - \psi|}, \quad pos_{end} = pos_s + \frac{(t_{end} - t_s) * |\vec{v}_s|}{r.length}$$

Next, let's consider how the NDTR-Tree is constructed and dynamically maintained through location updates. When the NDTR-Tree is first constructed in the moving objects database, the system has to read the route records of the traffic network and build the upper R-Tree. At this moment, all lower R-Trees are empty trees. After the construction, whenever a new location update message is received from a moving object, the server will generate corresponding trajectory units and insert them into the related lower R-Tree(s). Since active trajectory units contain prediction information, when a new location update occurs, the current active trajectory unit should be replaced by newly generated trajectory units.

Let's consider the situation when DTTLU or STTLU occurs. From the location update strategies for network constrained moving objects, we know that when DTTLU/STTLU happens, the moving object $mo$ is still running in route($rid_n$), where $rid_n$ is the route identifier contained in the current active motion vector $mv_n$. When the server receives a new location update message $mv_a = (t_a, rid_a, pos_a, \vec{v}_a)$ (where $rid_a = rid_n$), if $mv_a$ is the first motion vector of the moving object in RTree$_{low}$($rid_a$), then the system only need to insert $\mu(mv_a)$ into RTree$_{low}$($rid_a$) directly. Otherwise, the system has to do the following with RTree$_{low}$($rid_a$): (1) Delete $\mu(mv_n)$ generated at the last location update; (2) Insert $\mu(mv_n, mv_a)$ into RTree$_{low}$($rid_a$); (3) Insert $\mu(mv_a)$ into RTree$_{low}$($rid_a$). Figure 2 illustrates this process.

When IDTLU occurs, the process is relatively more complicated. Suppose that the moving object $mo$ transfers from route $r_s$ to $r_e$ (with route identifiers $rid_s$, $rid_e$ respectively). In this case three location update messages will be generated: $mv_{a1}, mv_{a2}, mv_{a3}$, where $mv_{a1}$ corresponds to the junction's position in $r_s$, $mv_{a2}$ corresponds to the junction's position in $r_e$, and $mv_{a3}$ corresponds to the location update position in $r_e$. The NDTR-Tree needs to make the following operations: (1) Delete $\mu(mv_n)$ from RTree$_{low}$($rid_s$), and insert $\mu(mv_n, mv_{a1})$ into RTree$_{low}$($rid_s$); (2) Insert $\mu(mv_{a2}, mv_{a3})$ into RTree$_{low}$($rid_e$); (3) Insert $\mu(mv_{a3})$ into RTree$_{low}$($rid_e$).

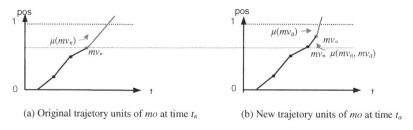

(a) Original trajetory units of *mo* at time $t_n$          (b) New trajetory units of *mo* at time $t_a$

**Fig. 2.** Maintenance of Lower R-Tree Records when DTTLU or STTLU Occurs

The constructing and dynamic maintaining algorithm for the NDTR-Tree is given in Algorithm 1. In the algorithm, the function mbr() returns the MBR of the given trajectory unit, and the functions Insert() and Delete() finish the insertion and deletion of trajectory units in the corresponding lower R-Trees respectively.

---

**Algorithm 1.** Constructing and Dynamic Maintaining Algorithm of NDTR-Tree

General Arguments:

$N = (Routes, Juncts)$;                    //the traffic network

1.   Read route records of $N$, and insert the related edges into the upper R-Tree;
2.   Set all lower R-Trees to empty tree;
3.   **While** (MOD is running) **Do**
4.      Receive location update package LUM from moving objects (Suppose the mov-obj ID is *mid*);
5.      **If** (LUM contains 1 motion vector $mv_a = (t_a, rid_a, pos_a, \vec{v}_a)$) **Then** // DTTLU or STTLU
6.         Let $mv_n$ be the current active motion vector of *mo* in $RTree_{low}(rid_a)$;
7.         **If** ($mv_n$ = NULL) **Then**
8.            **Insert**($RTree_{low}(rid_a)$, $(mid, \mu(mv_n)$, **mbr**$(mv_n)$));
9.         **Else**
10.           **Delete**($RTree_{low}(rid_a)$, $(mid, \mu(mv_n)$, **mbr**$(mv_n)$));
11.           **Insert**($RTree_{low}(rid_a)$, $(mid, \mu(mv_n, mv_a)$, **mbr**$(mv_n, mv_a)$));
12.           **Insert**($RTree_{low}(rid_a)$, $(mid, \mu(mv_a)$, **mbr**$(mv_a)$));
13.        **Endif**;
14.     **Else If** (LUM contains 3 motion vectors $mv_{a1}, mv_{a2}, mv_{a3}$) **Then** // IDTLU
15.        Let $mv_n$ be the current active motion vector of *mo* in $RTree_{low}(rid_{a1})$;
16.           **Delete**($RTree_{low}(rid_{a1})$, $(mid, \mu(mv_n)$, **mbr**$(mv_n)$));
17.           **Insert**($RTree_{low}(rid_{a1})$, $(mid, \mu(mv_n, mv_{a1})$, **mbr**$(mv_n, mv_{a1})$));
18.           **Insert**($RTree_{low}(rid_{a2})$, $(mid, \mu(mv_{a2}, mv_{a3})$, **mbr**$(mv_{a2}, mv_{a3})$));
19.           **Insert**($RTree_{low}(rid_{a2})$, $(mid, \mu(mv_{a3})$, **mbr**$(mv_{a3})$));
20.        **Endif**;
21.     **Endif**;
22. **Endwhile**.

---

Since in moving objects databases, the most common query operators, such as **inside** (*trajectory, Ix×Iy×It*), **intersect**(*trajectory, Ix×Iy×It*) (where *Ix, Iy, It* are intervals in X, Y, T domains), belong to range queries, that is, the input of the query is a range in the $X×Y×T$ space, we take range query as an example to show how the query processing is supported by the NDTR-Tree.

The range query processing on NDTR-Tree can be finished in two steps (suppose the range is *Ix×Iy×It*). The system will first query the upper R-Tree of the NDTR-Tree according to *Ix×Iy*, and will receive a set of (*rid × period*) pairs as the result, where *period* $\subseteq$ [0, 1] and can have multiple elements; then for each (*rid × period*) pair,

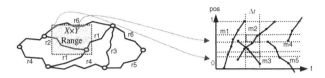

**Fig. 3.** Range Query through NDTR-Tree

search the corresponding lower R-Tree to find the trajectory units intersecting *period×It*, and output the corresponding moving object identifiers, as shown in Figure 3 and Algorithm 2.

---

**Algorithm 2.** Range Query Algorithm of NDTR-Tree

---

INPUT: *Ix×Iy×It;  //the querying Range*
OUTPUT: Result*;   //Set of moving object identifiers*

1.    Search the upper R-Tree according to *Ix×Iy*, and receive a set of pairs: $(rid_i, period_i)_{i=1}^{n}$ ;
2.    **For** ($i := 1$ to $n$) **Do**
3.      **For** $\forall \rho \in period_i \times It$ **Do**
4.        Let $\mu$ be the set of trajectory units in $RTree_{low}(rid_i)$ which intersect $\rho$;
5.        *Result= Result* $\cup$ the set of moving object IDs contained in the element of $\mu$;
6.      **Endfor**;
7.    **Endfor**;
8.    **Return** Result.

---

## 4   Implementation Issues and Performance Evaluation

The above stated RMODTN model has been implemented as a prototype within the PostgreSQL 8.2.3 extensible database system (with PostGIS 1.2.1 extension for spatial support), running on an IBM x205 with 17G hard disk, 256M memory, and Fedora Core 4 Linux operating system.

To evaluate the query processing performance of the RMODTN model, we have conducted a series of experiments based on the prototype system and some additionally implemented modules. The experiments are conducted based on the real GIS data of Beijing for the traffic network. In order to test the performance of RMODTN on different map scales, we generate three data sets with 4197, 9000, and 13987 routes respectively by selecting different route levels. Besides, we have implemented a network constrained moving objects generator, NMO-Generator, which can simulate moving objects according to predefined traffic networks.

In the experiments, we have first tested the query response time. We choose query examples 1 and 2 of Subsection 2.3 to test the query performance for region queries (to search moving objects inside a region) and for join queries (to search moving object pairs). In order to test the "pure" query processing performances, we do not utilize any index structures for moving objects in this step, even though traffic networks are indexed with B-tree (on *jid* and *rid*) and R-tree (on *geo*) respectively. The experimental results are shown in Figure 4.

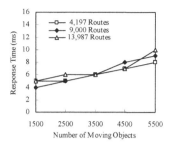
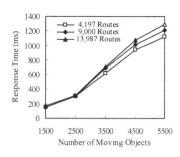

(a) Average Response Time for Region Queries    (b) Average Response Time for Join Queries

**Fig. 4.** Average Query Response Time of the RMODTN Model

From Figure 4(a) we can see that region queries can be processed efficiently in the RMODTN model. When the number of moving objects reaches 5500, the average response time is around 10ms, which is quite acceptable for typical querying users. Besides, with the map scale increasing, the query response time only increases slightly. From Figure 4(b) we can see that the response time for join queries increases when the number of moving objects going up. Nevertheless, the response time is still acceptable when the number of moving objects amounts to 5500.

In the experiments, we have also tested the performance of the proposed index methods. We take MON-Tree [1] as the control, and the results are shown in Figure 5.

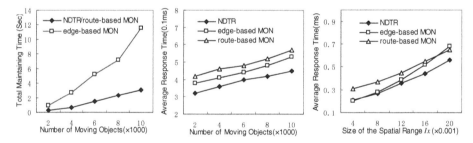

**Fig. 5.** Performance Comparison of UTR-Tree and MON-Tree

From Figure 5 we can see that the NDTR-Tree has better performances compared with the MON-Tree in dealing with the full dynamic trajectories of moving objects. This is because the NDTR-Tree utilizes a hybrid index structure. On the one hand, the upper R-Tree takes edges as the basic unit for indexing, which has smaller granularity so that the intersection between different MBRs can be greatly reduced. On the other hand, each lower R-Tree corresponds to a route, which has bigger granularity than edges, so that index maintaining costs can be reduced.

## 5   Conclusions

In this paper, a new moving objects database model, RMODTN, is proposed. In RMODTN, transportation networks are modeled as dynamic graphs and moving objects are modeled as moving graph points. Besides, a new index mechanism is proposed, and the implementation and performance evaluation results are discussed. Compared with other moving object models, RMODTN has the following features:

(1) The system is enabled to support logic road names, while queries based on Euclidean space can also be supported;

(2) Both history and current location information can be queried, and the system can also support near future location queries based on the predicted information;

(3) General events of the system, such as traffic parameters, blockages and topology changes can also be expressed so that the system is enabled to deal with the interaction between the moving objects and the underlying transportation networks.

As future research, we will deal with OLAP and data mining techniques based on moving object trajectories.

**Acknowledgments.** The work was partially supported by NSFC under grand number 60573164, and by SRF for ROCS, SEM. The author would like to thank Prof. Ralf Hartmut Güting of Fernuniversität in Hagen, Germany for his valuable discussions and advices on the first version of this paper.

# References

1. Almeida, V., Güting, R.H.: Indexing the Trajectories of Moving Objects in Networks. GeoInformatica 9(1), 33–60 (2005)
2. Ding, Z., Güting, R.H.: Modeling Temporally Variable Transportation Networks. In: Lee, Y., Li, J., Whang, K.-Y., Lee, D. (eds.) DASFAA 2004. LNCS, vol. 2973. Springer, Heidelberg (2004)
3. Ding, Z., Güting, R.H.: Managing Moving Objects on Dynamic Transportation Networks. In: Proc. of SSDBM 2004, Santorini, Greece (2004)
4. Ding, Z., Zhou, X.F.: Location Update Strategies for Network-Constrained Moving Objects. In: Haritsa, J.R., Kotagiri, R., Pudi, V. (eds.) DASFAA 2008. LNCS, vol. 4947. Springer, Heidelberg (2008)
5. Frentzos, E.: Indexing objects moving on fixed networks. In: Hadzilacos, T., Manolopoulos, Y., Roddick, J.F., Theodoridis, Y. (eds.) SSTD 2003. LNCS, vol. 2750. Springer, Heidelberg (2003)
6. Güting, R.H., Almeida, V.T., Ding, Z.: Modeling and Querying Moving Objects in Networks. VLDB Journal 2006 15(2)
7. Güting, R.H.: Second-Order Signature: A Tool for Specifying Data Models, Query Processing, and Optimization. In: Proc. ACM SIGMOD Conference, Washington, USA (1993)
8. Güting, R.H., Almeida, V., Ansorge, D., Behr, T., Ding, Z., et al.: SECONDO: An Extensible DBMS Platform for Research Prototyping and Teaching. In: Proc. of ICDE 2005 (2005)
9. Güting, R.H., Böhlen, M.H., Erwig, M., Jensen, C.S., et al.: A Foundation for Representing and Querying Moving Objects. ACM Transactions on Database Systems 25(1) (2000)
10. Papadias, D., Zhang, J., Mamoulis, N., Tao, Y.: Query processing in spatial network databases. In: Proc. of VLDB 2003, Berlin, Germany, (2003)

11. Pfoser, D., Jensen, C.S., Theodoridis, Y.: Novel Approach to the Indexing of Moving Object Trajectories. In: Proc. of the 26th VLDB, Cairo, Egypt (2000)
12. Sistla, A.P., Wolfson, O., Chamberlain, S., Dao, S.: Modeling and querying Moving Objects. In: Proc. of ICDE 1997, Birmingham, UK (1997)
13. Speicys, L., Jensen, C.S., Kligys, A.: Computational data modeling for network-constrained moving objects. In: Proc. of GIS 2003, Louisiana, USA (2003)
14. Su, J., Xu, H., Ibarra, O.: Moving Objects: Logical Relationships and Queries. In: Jensen, C.S., Schneider, M., Seeger, B., Tsotras, V.J. (eds.) SSTD 2001. LNCS, vol. 2121. Springer, Heidelberg (2001)
15. Vazirgiannis, M., Wolfson, O.: A Spatiotemporal Query Language for Moving Objects on Road Networks. In: Jensen, C.S., Schneider, M., Seeger, B., Tsotras, V.J. (eds.) SSTD 2001. LNCS, vol. 2121. Springer, Heidelberg (2001)
16. Wolfson, O., Xu, B., Chamberlain, S., Jiang, L.: Moving Object Databases: Issues and Solutions. In: Proc. of the 10th SSDBM, Capri, Italy (July 1998)

# Approach to Detection of Community's Consensus and Interest

Xijin Tang

Institute of Systems Science, Academy of Mathematics and Systems Science
Chinese Academy of Sciences, Beijing 100190 P.R. China
xjtang@amss.ac.cn

**Abstract.** Nowadays as Internet enables to find, publish and then share information among unfamiliar people and then enable virtual community emerges, it is natural to detect the consensus or interests from those on-line opinions or surveys, especially for those business people to acquire feedback and get senses which are beneficial for new prototypes design and products improvements. In this paper, several ways to approach community's consensus and interest are addressed. Those ways mainly denote three kind of technologies, augmented information support (AIS), CorMap and iView, which mainly support different kinds of work during an unstructured problem solving process where creative ideas are barely required.

**Keywords:** AIS, CorMap, iView, qualitative meta-synthesis.

## 1 Introduction

Internet creates a giant knowledge system. Currently if we want to know something, a search engine, e.g. *google* can pull many relevant or irrelevant web pages. We can also browse *wikipedia* to get more detailed information. E-commerce and e-business changes our traditional life style. Before buying something, we may go to search news about what we are concerned and may get many know-how replies from communities at BBS or some on-line shopping sites. Even we find partners to buy favorite things under a group purchase price, much lower than that at retail stores. Before traveling, Internet searching even becomes a necessary step and may even play important role to make a trip schedule. Business organizations also make use of the textual information to get some information about the feedbacks which are beneficial for new prototypes design and products improvements. That is one of driving forces of tide of business intelligence and data mining technologies.

As companies issue new products or services, customer survey is a usual way to get assessment. However people are not always patient with questionnaires, even during a so-called 5-minute phone-call investigation. They may scratch a few lines at the blank area at the answer sheet or post their ideas at a familiar on-line forum or BBS, instead of comparing those options directly. People may not show their attitude exactly during direct investigation. That is why to adopt new ideas toward finding customers' interest and opinions. In this paper, we address several ways to approach

Y. Ishikawa et al. (Eds.): APWeb 2008 Workshops, LNCS 4977, pp. 17–29, 2008.

community's consensus and interest. Those ways could be a supplement to those traditional approaches to potential new product designs or trends of customers' interests. The prediction of customers' preferences is usually an unstructured problem since always changing customers' appetites lead to many uncertainties. The following addressed technologies, augmented information support (AIS), CorMap and iView, mainly support different kinds of work in community opinion processing, are helpful to acquire a knowledge vision of unstructured problems, a result of qualitative meta-synthesis of community intelligence.

## 2  Knowledge Vision by Qualitative Meta-synthesis

During the unstructured problem solving process, we need to depict the problem in structured way so as to deal with it using known methods or their integrative ways. How to get some structures towards those problems is of more difficulties. It is necessary to get to know problem structuring methods so as to develop appropriate information technologies to help human information processing and decision making along the unstructured problems solving process.

### 2.1  Problem Structuring Approaches and Computerized Support

Analytical decision methods explain how to make choices among a set of alternatives. Oriented to substantive rationality, those computerized methods help to fulfill the third phase, i.e. *choice* within the Simon's intelligence-design-choice model of decision making. However, alternatives should be available at first. More attentions are required to be paid to the *intelligence* and *design* phases, where relevant tasks are undertaken through a problem structuring process. Lots of approaches to problem structuring are proposed mainly in Europe, especially in UK [1-5]. The *Wisdom* approach proposed by soft OR group at Lancaster University aims to procedural decision support [6]. A *Wisdom* process refers to facilitated session includes brainstorming, cognitive mapping and dialogue mapping. The cognitive mapping phase provides a macro view of the problem discussed by the group and the dialog mapping phase helps the group to develop consistent micro views. In parallel to many western schools in approaches and methodologies for unstructured problem solving, a Chinese system scientist Qian Xuesen (Tsien HsueShen) and his colleagues proposed meta-synthesis system approach (MSA) to tackle with open complex giant system (OCGS) problems from the view of systems in 1990 [7]. OCGS problems are usually regarded as unstructured problems. The essential idea of MSA can be simplified as from confident qualitative hypothesis to rigorous quantitative validation, i.e. quantitative knowledge arises from qualitative understanding, which reflects a general process of knowing and doing in epistemology. OCGS problem solving process goes through three types of meta-synthesis, (i) qualitative meta-synthesis; (ii) qualitative-quantitative meta-synthesis; and (iii) meta-synthesis from qualitative knowledge (hypotheses) to quantitative validation based on systems engineering practice [8]. The 1st type, qualitative meta-synthesis, aims to produce assumptions or hypotheses about the unstructured problems, i.e. to expose some qualitative relations or structures of the concerned problems. Computerized tools, such as group support systems (GSS), creativity support systems (CSS) etc. may support qualitative meta-synthesis. The working process of the qualitative meta-synthesis may be achieved by those problem structuring methods, such as the *Wisdom* approach, which may also support the third type of meta-synthesis to achieve final validated knowledge via facilitated collective intelligence. Different meta-synthesis needs different supports, which are of comprehensive discussions in correspondence with knowledge creating process in [9].

## 2.2  Qualitative Meta-synthesis to Knowledge Vision

MSA expects to take the advantages of both human beings in qualitative intelligence and machine system in quantitative intelligence to generate more new validated knowledge which may be stored into a conceptual knowledge system. The attention to the qualitative intelligence reflects the emphasis of human's dominant role at problem structuring and solving process, where resolutions about unstructured problems are captured through a series of structured approximation. For unknown or new issues, new ideas are often needed. Those new ideas may come from human's imaginary thinking, intuition and insight. Supported by creativity software, sparkling ideas may drop into one's mind. Creative solutions are often related with wisdom. Then, the practice of MSA is expected to enable knowledge creation and wisdom emergence.

At this point, we can sense that Internet is such a giant knowledge system, whose knowledge comes from contributions of Internet users facilitated by those various computer technologies enabled by the Web. A seraching engine can provide many *urls*. However people still need many efforts to acquire a rough vision of the concerned matter even with specialized supporting tools. If the interesting topic is across multiple disciplines, more efforts have to be taken to investigate those searching results. How to gain some senses about the concerned issue both efficiently and effectively is a problem. If got some hints or senses from those information, scenarios would then be drawn for further validating process, which may finally lead to a comprehensive knowledge vision of the issue. Thus a rough knowledge vision is required at first.

Then creative ideas are barely required than analytical or logical thinking. Creative thinking methods, such as brainstorming, KJ method, etc. are practical ways to acquire creative ideas, especially undertaken at group working level. Even complaints never fade toward low efficiency of group meeting, whatever is both feasible and effective for communication and information sharing, opinion collection and acquisition of expert knowledge. The Internet provides ideal context for active interactions, especially those empathic feedbacks and critical comments, and inevitably becomes a natural *ba* to harness the collective knowledge and creativity of a diverse community during collective problem solving process. The problem is how to faciliate such a process of acquiring knowledge vision from on-line community opinions, where augmented information technologies are being developed to provide somewhat helps.

## 3  Augmented Information Technologies toward Community's Consensus and Interests

Shneiderman abstracted four activities, collect, relate, create and donate for a framework of creativity based on creativity models [10]. The augmented information technologies are mainly developed to facilitate those four basic activities to acquire basic threads or constructs from those on-line community opinions, such as customers' preferences.

### 3.1  AIS – Augmented Information Support

The AIS technology actually refers to a large category of those widely studied Web and text mining technologies. Web text mining technology focuses the Web pages which contain not only pure texts but also the hyperlinks between Web pages and more preprocess should be counducted on Web texts than pure texts. Figure 1 shows a basic Chinese Web text mining process [11].

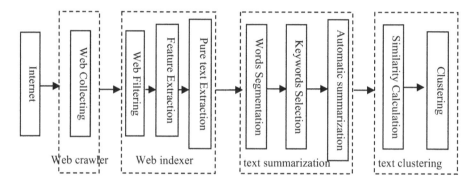

**Fig. 1.** A basic Chinese Web Text mining process [11]

All those technologies referred at Fig.1 are oriented to information collection, the 1st basic activity at the problem solving process or *intelligence* phase.

Besides the efficiency enhancement of searching, it is expected to acquire a set of records based on AIS technology with the structure as

&lt;topic, userID, text, keywords, time&gt;

Such metadata indicates the corresponding *userID* drops one *text* (e.g. one comment, one blog, a title of a paper, a reply to one question) with a set of *keywords* under the *topic* at the point of *time*. By word segmentation and filtered feature keywords used in text summarization, or even manual selection, a variety of human ideas and opinions can be transferred into one piece record of this form. The keywords for a blog may denote the tags. The keywords are articulated as attributes of the *userID* or the *text*. The next addressing technologies are based on such metadata.

### 3.2  Knowledge Vision by CorMap Analysis

CorMap analysis denotes a technology of exploratory analysis of textual data. Two major tasks will be carried out when CorMap analysis is applied.

### 3.2.1 Correspondence Analysis of Textual Data and Visualization of Correspondence Results

This technology mainly adopts correspondence analysis which provides a method of factoring categorical variables and displaying them in a property space where to map their association in 2 or more dimensions. This method has been widely used in many disciplines [12]. The singular value decomposition (SVD) is the principal mathematics of CorMap analysis. Given an $m$-by-$n$ matrix $Z$, SVD is defined as $Z = U\Sigma V^T$ where $U$ is an $m$-by-$m$ unitary matrix contains a set of orthonormal vectors called row singular vectors, $\Sigma$ is $m$-by-$n$ matrix where the diagnoal elements are nonnegative singular values sorted in descending order, and $V^T$ denotes the conjugate transpose of $V$, an $n$-by-$n$ unitary matrix contains a set of orthonormal vectors called column singular vectors.

By AIS technology, a dataset of community opinions with the quintuplet about one topic can be acquired and two contingency tables are formulated. Each table refers to a frequency matrix. The element of one matrix denotes the frequency of keyword $i$ referred by the participant $j$ relevant to the topic. The element of another matrix denotes the frequency of keyword $i$ referred by the text $j, i = 1, 2, \ldots, m$, $j = 1, 2, \ldots, n$. Then correspondence analysis is applied to both matrices and brings out two visual maps. By performing a series of transformations and SVD towards the transformed matrix, a set of row vectors and column vectors are achieved and then rescaled with the original total frequencies to obtain optimal scores. These optimal scores are weighted by the square root of the singular values and become the coordinates of the points. Given the coordinates, both participants and keywords can be mapped into 2-dimensional space. As a result, a pair of participants with more shared keywords may locate closer in the 2D space.

Such kind of analysis can be applied to any combination of available participants, and may help to "drill down" into those community thoughts to detect some existing or emerging micro community. If applied to an individual participant, CorMap analysis may unravel personal thinking structure.

Fig. 2 shows a CorMap of an investigation of working definition of knowledge science contributed by 20 faculty members and researchers of School of Knowledge Science at Japan Advanced Institute of Science and Technology (JAIST) taken in the end of 2006. Those respondents only wrote their understandings about what knowledge science research and education should be. Obviously differences existed among their understandings. If more than 2 sentences exist in one reply, more records for each reply are generated for better understanding. Then the whole data set includes 33 comments with a total of 74 filtered keywords.

CorMap aims to show a global thinking structure contributed by comments contributors, while different foci toward the topic could be easily seen. The respondent who locates close to the bottom of the visual map (Fig.2) obviously holds opinions far away from the majority.

Moreover three indicators, dominance, agreement and discrepancy, are provided to measure the contributions or roles of those participants relevant to the discussion of the topic. Who is active in posting comments? Who always follows others' ideas?

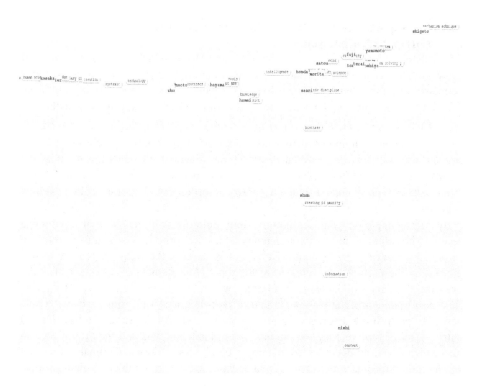

**Fig. 2.** CorMap visualization of a community towards working definition of Knowledge Science (20 People with 33 comments, 74 filtered keywords)

Who is of so peculiar thinking in this topic? Some of indicators were proposed in 2004 [13] and an improvement was taken in 2006 [14]. Such kind of work may also be helpful to select appropriate people for further study later.

### 3.2.2  Clustering of Ideas Based on Spatial Relations

Given the spatial relations acquired by correspondence analysis, a variety of clustering methods such as $k$-means clustering can then be applied to ideas clustering and concept extraction for qualitative meta-synthesis.

As shown in Fig.3, those 74-keyword spread across the 2D map is clustered into 6 clusters by $k$-means method. The keyword (whose label is of bigger size of fonts) which is closest to the centroid of the affiliated cluster could be regarded as the label of the cluster. The pop-up window lists all keywords and their frequencies in the Cluster labeled as "interdisciplinary" close to the left border of the map.

Furthermore instead of traditional human's judgment of clustering numbers by trial-and-error, CorMap analysis applies a method to determine the true number of clusters. This method is based on distortion, a measure of within cluster dispersion, which is from the field of rate distortion theory in the information theory [15].

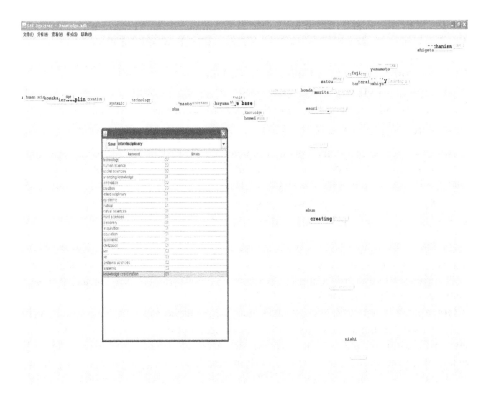

**Fig. 3.** Clustering of keywords (6 Clusters) and one pop-up box showing the keywords belonging to the left cluster labeled as *interdisciplinary*)

The motive to apply correspondence mapping about participants' opinions is to provide a basic or possible association between participants and their thoughts represented by keywords. As correspondence analysis is a method for exploratory analysis, significance testing is not supported. Then the visualized association is not confirmatory, even two dimensions may not visualize more than 75% of the association between humans and keywords. Therefore it is necessary to do further analysis instead of drawing conclusions directly from the visualized relevance. During the community opinion aggregating process, the dynamic mapping is to stimulate active association and feedback as a catalyst for shared understanding and wider thinking. A spontaneous and free-flowing divergent thinking mode is expected and possible helps are pushed for participants' awareness, even those hints are not confirmatory. Wild ideas toward the dynamic relevance, especially those isolated ideas far away from the majority may lead to some in-depth investigation for curiosity.

### 3.3 Knowledge Vision by iView Analysis

Given the same dataset used by CorMap analysis, iView analysis mainly applies network analysis methods to depict the scenario of the topic. The first step of iView analysis is to construct a pair of networks, one is keyword network (or text network), the other is human network. Here keyword network is addressed for illustration.

### 3.3.1  Keyword Network: A Collective Idea Map

In a keyword network $G = (K, E)$, the vertex refers to a keyword. If keyword $k_i$ and keyword $k_j$ simultaneously belong to the keyword set of one *text*, then an edge exists between two vertices $e_{ij} = (k_i, k_j), i \neq j$, $e_{ij} \in E$ ($E$ is the edge set). Then each keyword set of one *text* constructs a complete keyword graph, a somewhat structure of the text which reflect personal opinions. A keyword network denotes the aggregation of all keyword graphs based on all texts relevant to the concerned topic. If $G_l = (K_l, E_l)$ indicates the keyword graph of the $l$th text where $K_l = \{ k_1^l, k_2^l, \cdots, k_n^l \}$ is the keyword set, $E_l$ is the edge set, then $G = (K, E)$ where $K = \cup K_l = \cup \{k_1^l, k_2^l, \cdots, k_n^l\}$, $E = \cup E_l = \cup \{e_{ij}\}$, $i, j = 1, 2, \ldots, m$, $i \neq j$. This topological network is a weighted undirected network where the weight of edge denotes the frequency of co-occurrence of keywords among all texts. Such a network is referred as an *idea map* contributed by all participants.

Given such a network, more senses may be obtained via a variety of network analysis by detecting some of its features, such as cutpoints, centrality of keywords, keyword clustering, etc. which may expose different perspectives of a collective vision of all the participants. The underlying mathematics applied is mainly from graph theory social network analysis (SNA) [16]. For example, a cutpoint (articulation point) of a graph is a vertex whose removal increases the number of connected component [17]; then the cutpoint keyword may unravel the real key ideas. So does the centrality analysis of the keyword vertex. With use of community structure detection methods, clustering of keywords may help to understand the major points from those keyword clusters easier instead only by frequencies of individual keywords.

Fig.4 shows the keyword network of the above-mentioned investigation of knowledge science. Three components and nine cutpoints, *business, human, information, knowledge, KM* (knowledge management*), processes, science, systemic* and *tools*, are detected. Table 1 lists the top 15 keywords in measure of degree centrality and betweenness centrality, respectively. Obviously, different methods lead to different results. All cutpoints are among the top 15 keywords of centrality of betweenness, but only 6 appear within the high rank list of centrality of degree. Whatever, those central keywords may reveal how the community understand the concept of knowledge science, and may represent their interests in research and education. Based on Newman-Girvan algorithm for community clustering [18], 4 clusters are founded (the maximum value of the modularity measure $Q = 0.6482$) at the largest component with 62 keywords. The community structure analysis of a keyword network may provide some further senses than centrality of keywords about the consensus of the KS community.

**Fig. 4.** Community idea map of working definition of knowledge science (Cutpoint: vertex in square; modularity measure $Q = 0.6482$)

**Table 1.** The top 15 keywords in measure of centrality of both degree and betweenness

| Rank | Keyword | Degree | Keyword | Betweenness |
|------|---------|--------|---------|-------------|
| 1 | *knowledge* | 38 | *knowledge* | 1337.17 |
| 2 | *human* | 21 | *human* | 470.83 |
| 3 | *processes* | 14 | *systemic* | 228.00 |
| 4 | *science* | 14 | *processes* | 186.00 |
| 5 | creation | 11 | *science* | 160.83 |
| 6 | technology | 11 | *business* | 118.00 |
| 7 | innovation | 10 | *information* | 60.00 |
| 8 | creativity | 9 | creativity | 54.67 |
| 9 | discovery | 9 | intelligence | 45.00 |
| 10 | acquisition | 9 | creation | 35.50 |
| 11 | education | 9 | *km* | 29.50 |
| 12 | specialist | 9 | ke | 29.50 |
| 13 | civilization | 9 | innovation | 29.00 |
| 14 | *business* | 8 | technology | 12.00 |
| 15 | *information* | 7 | *tools* | 5.00 |

### 3.3.2  Human Net Via Keyword-Sharing between Humans

In iView analysis, keywords-sharing between participants is considered and a human network where the vertex denotes a participant is constructed. If two participants share one keyword, a link between exists. The strength between two participants indicates the number of the different keywords or the total frequencies of all the keywords they share. From such a human net, social network analysis (SNA) is then applied to find the powerful people by centrality analysis and to detect the interest group by community structure detection, etc.

As shown in Fig.5, the keyword-sharing network of KS investigation has 1 component, which is split into 3 clusters. 2 cutpoints (*Wier* and *Shun*) are detected. The top 4 keywords in centrality of degree are *Saori* (15), *Morita* (14), *Honda* (13) and *Wier* (10). There are 6 vertices with a degree value of 6, *Shun* is among them. The top 5 keywords in centrality of betweenness are *Saori* (26.65), *Wier* (26.00), *Honda* (25.75), *Morita* (20.65) and *Shun*(18.00). All other humans' betweenness values are below 10.00. The pop-up window shows one respondent's reply (text, keyword and its frequency).

**Fig. 5.** Human net of KS  investigation (3 Clusters, $Q = 0.3245$)

Small scale discussion always brings out human net with one giant component. If the topic scope expands, more components may emerge.

The pair of idea map and human net could be regarded as one kind of structure about the dedicated topic. The exploited network analysis aims to detect basic concepts and main themes, influential people and the potential micro communities emerged during the discussion. Those information could be regarded as the constructs of the concerned topic and may be helpful to those latecomers or entrants to quickly get a rough understanding of the interesting topic, a consensus of the existing community.

We can also take iView analysis to text network according to the keyword reference. Obviously, the text network is a directed network. This network may help to show how the ideas grow and spread during the discussion. Due to space limitation, detailed discussions are omitted.

## 4  Concluding Remarks

In this paper, we address three kinds of augmented information technologies, AIS, CorMap and iView, which are oriented to detect community's consensus and interest from those opinions contributed at BBS or various on-line forums. AIS technology aims to collect more useful textual information from the Web. The goal of both Cor-Map and iView analysis is to depict the collective vision toward the concerned issue by different ways to explore relevance between the texts and their contributors, support awareness and insights for potential hints, ideas, or senses of problems from which to undertake quantitative modeling for alternatives during problem solving process. The information detected by those data-driven, textual computing based visual exploration technologies could be pushed to the people to stimulate active participation, to show a map of human or community thinking process and then to construct a holistic vision by community opinions.

It is easier for people to show real attitude due to the decentered subjectivity encouraged by the Internet which also enables the on-line community creation. Nowadays, emerging technologies are being explored and exploited to improve the effectiveness of information retrieval and understanding.  The tag cloud shows a visual depiction of user-generated tags used typically to describe the content of Web pages/sites. The cloud model lists the feature words alphabetically and the importance of a word is shown with font size or color. At this point, both CorMap and iView analytical technologies expose further relevance of those words and have it visualized.  Web 2.0 technologies enable human participation and provide context to apply CorMap and iView technologies to quickly get a rough understanding of the interesting topic. For new comers or bystanders, it is easier to know some points from the visualized maps of community thinking, find fancy ideas which may lead to some in-depth investigation for curiosity.

The three technologies had already been applied to a famous science forum in China [11, 19, 20]. Both CorMap and iView technologies had integrated into different kind of tools for different uses, such as TCM Master Miner which exhibits a special way of TCM master thoughts' mining [21]. The iView analysis is of more application in conference mining to show the collective knowledge vision of one discipline [22]. Others' ideas are worth adopting or comparing, such as chance discovery using Key-Graph [23] and skillMap [24] for further explorations of current technologies.

## Acknowledgement

This work is supported by Natural Sciences Foundation of China under Grant No. 70571078 & 70221001. The author is grateful to Mr. Kun NIE, a doctoral student at JAIST, who provided the original investigation data which was then processed and transferred to be as illustrations in this paper.

## References

1. Tomlinson, R., Kiss, I. (eds.): Rethinking the Process of Operational Research and System Analysis. Pergamon, Oxford (1984)
2. Flood, R.L., Jackson, M.C.: Creative Problem Solving: Total Systems Intervention. John Wiley & Sons, Chichester (1991)
3. Rosenhead, I., Mingers, J. (eds.): Rational Analysis for a Problematic World Revisited, 2nd edn. John Wiley & Sons, Chichester (2001)
4. Detombe, D.J. (ed.): Handling complex societal problems (special issue). European Journal of Operational Research 128(2), 227–458 (2001)
5. Slotte, S., Hämäläinen, R.P.: Decision Structuring Dialogue, Systems Analysis Laboratory, Helsinki University of Technology, Electronic Reports (2003), http://www.e-reports.sal.tkk.fi/pdf/E13.pdf
6. Mackenzie, A., et al.: Wisdom, decision support and paradigms of decision making. European Journal of Operational Research 170(1), 156–171 (2006)
7. Qian, X.S., Yu, J.Y., Dai, R.W.: A new discipline of science - the study of open complex giant systems and its methodology. Nature Magazine 13(1), 3–10 (1990); (in Chinese, an English translation is published in Journal of Systems Engineering & Electronics 4(2), 2–12(1993))
8. Yu, J.Y., Tu, Y.J.: Meta-Synthesis - Study of Case. Systems Engineering - Theory and Practice 22(5), 1–7 (2002) (in Chinese)
9. Tang, X.J.: Towards Meta-synthetic Support to Unstructured Problem Solving. International Journal of Information Technology & Decision Making 6(3), 491–508 (2007)
10. Shneiderman, B.: Creativity Support Tools. Communications of the ACM 45(10), 116–120 (2002)
11. Zhang, W., Tang, X.J., Yoshida, T.: Web Text Mining on a Scientific Forum. International Journal of Knowledge and Systems Sciences 3(4), 51–59 (2006)
12. Beh, E.J.: Simple Correspondence Analysis: a Bibliographic Review. International Statistical Review 72(2), 257–284 (2004)
13. Tang, X.J., Liu, Y.J.: Exploring Computerized Support for Group Argumentation for Idea Generation. In: Nakamori, Y., et al. (eds.) Proceedings of the 5th International Conference on Knowledge and Systems Sciences (KSS 2004), pp. 296–302. JAIST Press, Japan (2004)
14. Tang, X.J. (ed.): Meta-synthesis and Complex Systems (2006 -2007). Research Report No. MSKS-2007-05, Academy of Mathematics and Systems Science, Chinese Academy of Sciences, pp. 47-50 (2007) (in Chinese)
15. Sugar, C.A., James, G.M.: Finding the number of clusters in a dataset: an information-theoretic approach. Journal of the American Statistical Association 98(463), 750–763 (2003)
16. Hanneman, R.A., Riddle, M.: Introduction to Social Network Methods, University of California, Riverside (2005), http://faculty.ucr.edu/hanneman/nettext/

17. Harary, F.: Graph Theory, ch. 3. Addison-Wesley, Reading (1969)
18. Newman, M.E.J., Girvan, M.: Finding and Evaluating Community Structure in Networks. Physical Review E 69, 026113 (2004)
19. Tang, X.J., Liu, Y.J., Zhang, W.: Augmented Analytical Exploitation of a Scientific Forum. In: Iwata, S., et al. (eds.) Communications and Discoveries from Multidisciplinary Data. Studies in Computational Intelligence, vol. 123. Springer, Heidelberg (2008) (the first paper in chapter 3)
20. Zhang, W., Tang, X.J.: A Study on Web Clustering with respect to XiangShan Science Conference. In: Iwata, S., et al. (eds.) Communications and Discoveries from Multidisciplinary Data. Studies in Computational Intelligence, vol. 123. Springer, Heidelberg (2008) (the first paper in chapter 7)
21. Tang, X.J., Zhang, N., Wang, Z.: Exploration of TCM Masters Knowledge Mining. Journal of Systems Science and Complexity 21(1), 34–45 (2008)
22. Tang, X.J., Zhang, Z.W.: How Knowledge Science is Studied - a Vision from Conference Mining of the Relevant Knowledge Science Symposia. International Journal of Knowledge and Systems Sciences 4(4), 51–60 (2007)
23. Matsuo, Y., Ohsawa, Y., Ishizuka, M.: KeyWorld: Extracting Keywords from a Document as a Small World. In: Jantke, K.P., Shinohara, A. (eds.) DS 2001. LNCS (LNAI), vol. 2226, pp. 271–281. Springer, Heidelberg (2001)
24. Meyer, B., Spiekermann, S., Hertlein, M.: skillMap: Identification of parallel developments and of Communities of Practice in distributed organizations. In: Gu, J.F., Chroust, G. (eds.) Proceedings of the First World Congress of the International Federation for Systems Research (IFSR 2005), Kobe, Japan, November 14-17. JAIST Press (2005) No. 20053

# A Comparative Empirical Study on the Margin Setting of Stock Index Futures Calendar Spread Trading*

Haizhen Yang[1], Hongliang Yan[1], and Ni Peng[2]

[1] School of Management, Graduate University of Chinese Academy of Sciences,
Chinese Academy of Sciences, Beijing 100080, China
[2] CITIC International Contracting Inc.
CITIC International Contracting Inc, Beijing 100004, P.R. China
haizheny@gucas.ac.cn, yanhongliang05@mails.gucas.ac.cn,
pengni@citic.com

**Abstract.** Using the data of Hang Seng index futures, this empirical research investigates EWMA method, ARMA-EGARCH model and EVT in order to provide a scientifically prudent and practically available approach for the margin setting of calendar spread trading in China. The results show that the three models above have their own advantages and disadvantages, taking the stability and practical feasibility into consideration. EWMA is simple in calculation and easy to be put into practice, but it may lead to an underestimation of the market risk due to the inaccurate decay factor in the model. EVT is prudent, but not easy to be widely implemented due to the requirement for chronic data accumulation. ARMA-EGARCH has the best performance in both accuracy of risk estimation and feasibility of practical implementation. To a certain extent, the research work provides both theoretical and empirical support for the margin setting of the coming CSI300 index futures.

**Keywords:** Futures Calendar Spread Trading, Margin, EWMA, ARMA-EGARCH, EVT.

## 1 Introduction

The development of Futures Market requires a complete set of risk management system, in which the mechanism of margin setting plays a prominent role. Without doubt, an appropriate margin setting can help strengthen the market capability to withstand the risks and improve the market efficiency as well.

As a prevailing international practice, dynamic margin system allows margin levels to be adjusted according to some relevant factors such as trading purposes, fluctuation of futures price and transaction activities, etc. The margin level of futures calendar spread trading set by most famous international exchanges is obviously lower than the one of speculating transaction. Aside from the easy terms of margin, compared to correspondents in China, most exchanges ask for lower transaction cost and looser

---

* The research is supported by NSFC (grant No.: 70673100, 70621001) and Graduate University of Chinese Academy of Sciences.

Y. Ishikawa et al. (Eds.): APWeb 2008 Workshops, LNCS 4977, pp. 30–41, 2008.

position limit, which do help decrease the expense of spread trading and increase the transaction volume and liquidity in futures market. There are mainly two reasons for such an easy margin requirement in calendar spread trading business: First, basically speaking, since spread trading business makes the profit based on the spread changes between relative contracts, it bears lower market risk than speculating does. Second, calendar spread trading further strengthens the price linkage between relative contracts, which helps to enhance market pricing function and improve market efficiency. Obviously, an efficient market will in turn sustain a further reduced risk in spread trading.

Then we move to the current requirements in China. Up to now, China implements the static margin system in the Futures transaction business, under which a fixed margin level is required. Although it allows slight adjustment of the margin level according to two measuring factors (net position of the futures contracts and the different phases of contracts), it may still lead to an either too high or too low margin level, because the market risk can not be precisely predicted merely by the said measuring factors. Furthermore, most of the exchanges in China except Zhengzhou Commodity Exchange have a margin requirement for the position in futures calendar spread trading to both the buyer and the seller as futures calendar spread trading has been regarded as a kind of speculating. Compared to dynamic margin system, the static margin system surly increases the opportunity cost, decreases market liquidity and weakens the price linkage between relative contracts. From a long-term point of view, it will further weaken the market pricing function and reduce market efficiency.

As the Shanghai Shenzhen 300 index (CSI300) futures has been in process of preparation for launching by China Financial Futures Exchange (CFFEX), a more flexible mechanism of margin setting has turned out to be a pressing need in China. In the junior phase, it is advisable to learn from international successful experiences and build on relative theoretical innovations.

The studies on futures margin setting mainly concentrated on the sufficiency of margin setting under Extreme Value Theory (EVT) model. Longin (1999)'s study on the Optimal Margin Levels in Futures Markets under EVT is one of the representatives of the achievements in this field. By estimating the margin level of the silver futures contract in COMEX based on two premises respectively, his study showed that, comparing with non-parametric statistics, normal distribution would probably lead to a certain underestimation [1]. Another typical achievement in this field is presented by Cotter (2001). He applied the Hill estimator [2] under EVT to measure downside risk for European stock index futures markets, and confirmed that, the margin level set by those European exchanges was fairly sufficient to cover the risk they predicted [3].Meanwhile, the studies in China mainly focused on the margin setting estimation of speculating transactions, which is based on the fluctuation of index spot price[4], relatively little attention has been paid to the margin setting on stock index futures. Thus it still leaves an open space for comparative empirical research in the practical issue-an appropriate margin setting of Chinese stock index futures. Therefore, the aim of this paper is to make contribution in this very subject. Using the data of Hang Seng index futures, this research selects three of most dominating methods for empirical analysis to find a scientifically correct and practically available approach for the margin setting of the coming CSI300 index futures transaction. What's more, we hope it can also provide a further theoretical and empirical support for the establishment of related mechanism of margin setting of Chinese stock index futures.

The paper is organized as follows. Section 2 describes the current prevailing methods for margin estimation and the main principles and hypothetical premises for each. Section 3 introduces the data used, examines different methods and models for margin estimation and makes a comparison. Section 4 presents conclusions including suggestions that vary with different periods of market development.

## 2   Types of Methods for Margin Setting of Stock Index Futures Calendar Spread Trading

International derivatives markets are operating on a mature system of margin setting. Their core principle is to estimate risks and determine margin level mainly through risk assessments. There are various types of popular methods based on this rule. EWMA method and GARCH family models are generally accepted as well and have been adopted by most exchanges. The method based on EVT is another popular focus in academic discussion.

### 2.1  EWMA

Exponentially Weighted Moving Average (EWMA) assigns a greater weight to those more recent observations, making the predicted volatility more sensitive to recent variance of asset prices. It is a method that can explain volatility clustering, and produce higher precision.

The result of EWMA is as follows:

$$\overset{\Lambda}{\mu}_t = \frac{\sum\limits_{i=1}^{n} \lambda^{i-1} \Delta F_{i-1}}{\sum\limits_{i=1}^{n} \lambda^{i-1}} \tag{1}$$

$$\overline{\sigma}_t = \sqrt{\frac{\sum\limits_{i=1}^{n} \lambda^{i-1} \times (\Delta F_{i-1} - \overset{\Lambda}{u}_t)}{\sum\limits_{i=1}^{n} \lambda^{i-1}}} \tag{2}$$

Where $\overset{\Lambda}{\mu}_t$ represents predicted mean of the underlying instrument, $\sigma_t$ is predicted volatility of the underlying instrument, $\Delta F_k$ stands for the change of the underlying instrument on the $k$ th trading day, $\lambda$ is decay factor of the EWMA method.

The major weakness of EWMA is that, there is no mature and stable approach to estimate the value of the decay factor at present. The optimal approach has been explored for decades and the effort still keeps going on. Hong Kong stock exchange chooses fixed decay factor ($\lambda = 0.96$) in practical application. Fan Ying (2001) employed root mean squared error (RMSE) to determine the decay factor [5]. Liu Yifang, Chi Guotai etc. YU Fangping , SUN Shaohong and WANG Yu-gang (2006) more precisely estimated the decay factor using lag coefficient in GARCH model [6].

## 2.2  GARCH Family Models

In order to obtain a more precise estimation of volatility, GARCH family models have been applied to the prediction of volatility, with which the maintenance margin can be better estimated.

Bollerslev (1989) developed GARCH model [7], overcoming to some extent the shortcomings of redundant parameters and difficulties of estimation found in the former ARCH model by Engle (1982) [8]. Furthermore, GARCH can describe volatility clustering and partly explain high kurtosis and heavy-tail.

Subsequently, the Exponential GARCH (EGARCH) model, which can capture asymmetry noise and better portray good news and bad news's asymmetric impact on volatility [9], had been brought forward by Nelson (1991), in hopes of solving the problem of level-effect of dynamitic assets. The general form of this model is:

$$\ln(h_t) = \omega + \sum_{i=1}^{q}\left[\alpha_i\left|\frac{\varepsilon_{t-i}}{\sqrt{h_{t-i}}}\right| + \phi_i\frac{\varepsilon_{t-i}}{\sqrt{h_{t-i}}}\right] + \sum_{j=1}^{p}\beta_j\ln(h_{t-j}) \tag{3}$$

Where $\varepsilon_t$ is residual term, $\varepsilon_t = \sqrt{h_t}\cdot\upsilon_t$, $\upsilon_t$ i.i.d, and E($\upsilon_t$)=0,D($\upsilon_t$)=1。$\omega$ is const, $\alpha_i$ is the return coefficient, $\beta_j$ is the lag coefficient, $\phi_i$ is the level-effect coefficient, whether $\phi_i$ is positive or negative decides the good news and bad news's asymmetric impact on volatility.

GARCH models are extensively used in empirical financial analyses. SGX-DC conducts volatility studies on a regular basis to ensure that the margin levels set are reflective of the changes in volatilities, incorporating the recent market movements.

## 2.3  Methods Based on EVT

Both EWMA method and GARCH models assume that distributions of financial returns follow certain particular types of distribution, in which the preference setting has been concerned with the over-all distribution of financial returns. However, the hypothesis may inaccurately estimate the tail risk, causing an under or over estimation of margin setting.

In fact, in assessing the risks, the tail of the financial returns distribution should be concentrated more. EVT does not assume any over-all distribution of the returns series. Instead, it only emphasizes the characteristics of the heavy tails. Well approaching the tail of financial returns distribution, EVT can provide a more precise margin level estimation than parametric statistical techniques.

The Generalized Pareto Distribution (GPD) is the limiting distribution of normalized excesses over a threshold, as the threshold approaches the endpoint of the variable (Pickands, 1975) [10]:

$$G_{\xi,\beta}(x) = \begin{cases} 1-(1+\xi x/\beta)^{-1/\xi} & \xi \neq 0 \\ 1-\exp(-x/\beta) & \xi = 0 \end{cases} \tag{4}$$

Where $\beta$ is the scale parameter, $\xi$ is the shape parameter which determines the shape of the distribution, and $\tau = \dfrac{1}{\xi}$ is the defined tail index of the distribution.

There are two major methods to estimate tail index: the conventional Hill estimation proposed by Pickands (1975) [11] and Hill (1975) [2] and $VaR - X$ model developed by Huisman, etc. (1998) [12].

$VaR - X$ Estimation assumes that the financial returns series are t-distribution. Based on the degrees of freedom (DF) of the t-distribution which is estimated according to the returns of the sample data, we can accurately estimate the tail distribution of the financial returns(See details in [12]). Given the confidence level $\alpha$, the margin level can be calculated as follows:

$$M = \sqrt{\frac{\delta}{\xi(\xi-2)}}\alpha + \mu \tag{5}$$

Due to complexity of calculation and strict requirement for chronic data accumulation on assets prices, the application of EVT in futures exchanges and clearing houses has met with certain difficulties. Despite of that, because of its accurate measuring of extreme risk, EVT has already been widely used in the field of reinsurance and measurement of credit risk, where its requirement for chronic data accumulation can be met.

## 3   Empirical Study on the Margin Setting of Hang Seng Index Futures Calendar Spread Trading

Hong Kong stock market is the most important oversea market for china-concept stocks, and it demonstrates a lot of similarities (e.g. a time limit on trading, trading mechanism) with domestic capital market. Therefore, we here adopt the data of Hang Seng index futures to investigate margin setting of futures calendar spread in this research work.

### 3.1   Research Technique and Data Description

The sample period analyzed in this paper ranges from Jan 25th, 2000, when HKCC has shifted to the present EWMA from Simple Moving Average (SMA), to May 4th, 2007.

There are four kinds of contracts trading in the stock market with different delivery periods (i.e. the spot month, the next calendar month, and the next two calendar quarterly months). Due to the expiration effect of spot month contract, the delivery risk may be additionally involved in the risk of futures calendar spread trading, then probably cause an irrational overestimation on transaction risk. Therefore, only the data of continuous contracts *HSIc3* and *HSIc4* [1] will be discussed in this testing. In addition, although a contract series may exhibit considerable jumps on the day of switching contracts. in order to preserve the continuity of the time series, we remain the data in such cases in the present time serial modeling. Inter-delivery spread is defined as :

$$\Delta F_t = HSIc4_t - HSIc3_t \tag{6}$$

Yield of inter-delivery spread（also called "spread yield"）is defined as :

$$R_t = (\Delta F_t - \Delta F_{t-1}) / S(t) \tag{7}$$

Where $\Delta F_t$ is the inter-delivery spread, $S(t)$ is the spot price of the underlying instrument. Because the margin level is determined by the fluctuation of the portfolio, the margin setting of futures calendar spread trading will be calculated on a basis of the yield of inter-delivery spread.

### 3.2  Hypothesis Testing

Spread yield of HSIF vibrates in the vicinity of zero and demonstrates apparent volatility clustering, which shows significant heteroscedastic phenomenon exists in the yield series while it is not a white noise process.

*(1) Correlation Test and Partial Correlation Test*

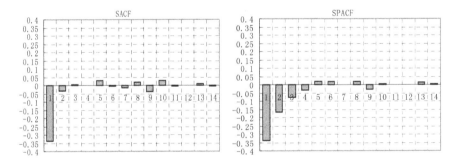

**Fig. 1.** Correlation test and partial correlation test for spread yield of Hang Seng index futures respectively

The mean of yield series holds high-correlation and then goes to zero immediately, which indicates that the yield series satisfies the requirement of stability. Besides, Fig.1 illustrates that it is an ARMA process. Therefore, in the process of time series modeling, ARMA model should be introduced while the yield series should satisfy ARMA (3, 1) process or AR (3) process.

---

[1] The data of continuous futures contracts ( *HSIc3* and *HSIc4* ) are sourced from Reuters 3000Xtra Terminal. *HSIc3* is the continuous futures contract which connects all the current 1st quarterly contracts in the series. And *HSIc4* is the continuous futures contract which connects all the current 2nd quarterly contracts in the series.

*(2)  Unit Root Test*

Table 1. Unit Root Test for spread yield of Hang Seng index futures

| Unit Root Test | -31.58483 | 1% critical value | -3.433863 |
|---|---|---|---|
| | | 5% critical value | -2.862978 |
| | | 10% critical value | -2.567583 |

The statistic for Unit Root Test is less than 1% critical value. At the 5% significance level, the test rejects the null hypothesis of a unit root in the spread yield series, which indicates that the time series of Hang Seng index futures' spread yield is a steady one. Based on the stationarity of spread yield, we can estimate the implied volatility exactly with the aid of EWMA and AR-EGARCH methods and then set prudent margin levels of futures calendar spread.

## 3.3  Results of Empirical Test

### 3.3.1  EWMA Method

According to the Hong Kong stock exchange, a decay factor $\lambda$ of 0.96, and a prediction window length of 90 days are used to estimate the margin level. By using Matlab 7.0 and EWMA method, the predicted mean $\hat{\mu}_t$ and predicted volatility $\sigma_t$ can be calculated.

Moreover, based on Normal Distribution assumption, the margin level of exchanges can cover 99.74% risk. And the margin level of brokers can cover 99.99% risk.

$$M_{Exchange} = \left| \hat{\mu}_t \right| + 3\sigma_t \tag{8}$$

$$M_{Broker} = \left| \hat{\mu}_t \right| + 4\sigma_t \tag{9}$$

The volatility $\sigma_t$ of the mean $\hat{\mu}_t$ can be calculated by EWMA method. Thereafter, the margin levels of exchanges and brokers can be estimated, as shown in Figure 2.

Based on EWMA method, the dynamic margin level can also be calculated. It reports that, in general, the margin level of futures calendar spread is significantly higher than the daily fluctuation of the spread yield. After eliminating the data on the days of switching contracts, we find that, at 99.99% confidence level, there are only 10 of 1703 trading days in the sample that have a daily price difference fluctuation that exceeds the margin level. Therefore, EWMA is an easy and effective method for margin setting estimation. The margin level set by Hong Kong stock exchange, which is based on EWMA method and, can basically cover the risk of calendar spread, and bring down default risk.

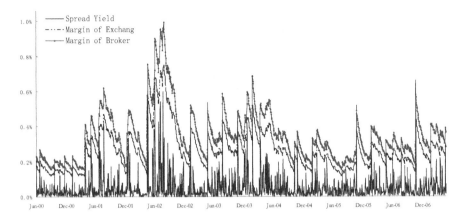

**Fig. 2.** Margin level calculated by EWMA method

### 3.3.2 ARMA-EGARCH Model

From correlation test and partial correlation test, we see that spread yield series shows a significant auto-correlation and it has been found to be a typical ARMA-GARCH series. According to AIC, SC and likelihood Estimation index, we adopt AR(3)-EGARCH(1,1) model to describe the movement of spread yield series. Estimating the parameters of AR(3)-EGARCH(1,1) model, we get that:

**Table 2.** The Parameter estimated by AR(3)-EGARCH (1, 1) model

| Parameter | Estimated Parameter | Volatility | p-value |
|:---:|:---:|:---:|:---:|
| AR(1) | -0.414063 | 0.028520 | 0.0000 |
| AR(2) | -0.201989 | 0.024515 | 0.0000 |
| AR(3) | -0.065353 | 0.021807 | 0.0027 |
| $\omega$ | -7.526127 | 0.461557 | 0.0000 |
| $\beta$ | 0.510041 | 0.026390 | 0.0000 |
| $\gamma$ | -0.009337 | 0.018655 | 0.6167 |
| $\alpha$ | 0.494856 | 0.031500 | 0.0000 |

EGARCH (1, 1) model can be described in the following

$$u_t = AR(1)u_{t-1} + AR(2)u_{t-2} + AR(3)u_{t-3} + \varepsilon_t \tag{10}$$

$$\delta_t^2 = \exp(\omega + \beta \ln \delta_{t-1}^2 + \alpha \left| \frac{\varepsilon_{t-1}}{\delta_{t-1}} - \sqrt{\frac{2}{\pi}} \right| + \gamma \frac{\varepsilon_{t-1}}{\delta_{t-1}}) \tag{11}$$

According to equations above, we get the forecasting of mean and volatility. Base on the quantile of normal distribution and estimated parameters in   equation (8) and equation (9), we can set the margin levels which rely on AR(3)-EGARCH(1,1) model.

### 3.3.3 EVT: Evaluating Extreme-Value Parameter Estimated by VaR-X Model

EVT can set the margin level more precisely in cases when there is not any exact assumption of distribution of spread yield. This paper evaluates extreme-value parameter-the tail index of a Pareto distribution-by VaR-X model. At 99.74% and 99.99% confidence level, the margin level of calendar spread trading can be estimated respectively by the said VaR-X model.

First, we sort the series of spread yield in ascending order, and then calculate $(\tau(k), k)$ sequences by equation $\tau(k) = \dfrac{1}{k} \sum\limits_{i=1}^{m} \left[ \ln R_{n+1-i} - \ln R_{n-k} \right]$. Using a linear regression $\tau(k) = \beta_0 + \beta_1 k + \varepsilon(k)$, we get the result: $\beta_0 = 0.33026$, $\beta_1 = 0.000765$.

The tail index of a Pareto distribution is determined as $\xi = \dfrac{1}{\tau} = \dfrac{1}{\beta_0} = 3$. By plugging it into equation $\tau(k) = \dfrac{1}{k} \sum\limits_{i=1}^{m} \left[ \ln R_{n+1-i} - \ln R_{n-k} \right]$, we get the estimation of the correspondent margin levels under EVT. At 99.74% confidence level, margin level is 0.427% and at 99.99% confidence level, margin level is 1.246%, both of which are significantly higher than the margin levels calculated based on EWMA method and ARMA-EGARCH model.

### 3.4 Comparative Analysis

Comparing the margin levels calculated by EWMA method, AR(3)-EGARCH(1,1) model and EVT, we can see that, they are all significantly higher than the yield of futures calendar spread trading, which can efficiently cover the market risk. The margin level estimated by AR(3)-EGARCH(1,1) model is slightly higher than that estimated by EWMA method.

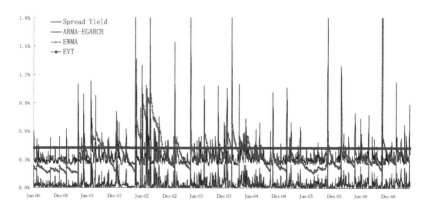

**Fig. 3.** Margin levels calculated by EWMA, ARMA-EGARCH and EVT

To examine the sufficiency, next, we carry out the Kupiec testing [13] on the above results. Table 4 and Table 5 illustrate the margin levels, probability of default, and LR-statistics. In Kupiec Test, the LR-statistics are 9.068854 under 99.74% and 15.13671 under 99.99% significance respectively. Here we see that, at the 99.74% confidence level, the margin level estimated by EWMA method is less than sufficient to pass the test. At the confidence level of 99.99%, the margin level calculated by EVT is far higher than twice the maximum in the samples, which implies that, EVT may overestimate the risk of the quarterly contracts calendar spread trading, causing a correspondently higher transaction cost.

**Table 3.** Default Probability of Margin setting estimated by EWMA, ARMA-EGARCH and EVT at the confidence level of 99.74%

| 99.74% confidence level | Margin level | Default days | probability of default | LR-statistics |
|---|---|---|---|---|
| EWMA | 0.278% | 42 | 2.466% | 49.80198 |
| ARMA-EGARCH | 0.289% | 16 | 0.94% | 7.836656 |
| EVT | 0.427% | 13 | 0.763% | 4.734741 |

**Table 4.** Default Probability of Margin setting estimated by EWMA, ARMA-EGARCH and EVT at the confidence level of 99.99%

| 99.99% confidence level | Margin level | Default days | probability of default | LR-statistics |
|---|---|---|---|---|
| EWMA | 0.366% | 10 | 5.872% | 2.244236 |
| ARMA-EGARCH | 0.375% | 7 | 4.11% | 0.552249 |
| EVT | 1.246% | 0 | 0% | 0 |

## 4  Concluding Remarks

An appropriate approach for margin setting in an emerging market (like that of China) remains highly concentrated and has been kept under academic discussion for years. In the search for clearer answers, the main theoretical methods that are currently popular ought to be tested. The main contribution of this paper is to select the data of Hang Seng index futures, which is highly similar to the features of domestic capital market among the abundant data storage, and isolate the spread yield of quarterly contracts for consideration, and then examine how EWMA method, ARMA-EGARCH model and EVT applied on the margin setting estimation of calendar spread trading. Thus, this paper is able to scrutinize their applicability respectively, and then identify suggestions in practice.

The results show that the three models above have their own advantages and disadvantages for comprehensive consideration of the stability and practical feasibility. In general, EVT will conduct an overestimation of the market risk under the premise of a small sample, which will probably cause a high transaction cost and a decrease in calendar spread trader's enthusiasm. Contrarily, EWMA method may cause events of default by conducting an underestimation of the market risk, since the fixed decay factor is not easy to be precisely calculated in the model. Among these three methods, ARMA-EGARCH model has the best performance in both accuracy of risk estimation and feasibility of practical implementation.

The suggestions for application in China vary with different phases of market cultivation and development. Considering that the retail investors possess a large percentage of Chinese investors at present, the price fluctuations may be extensively fierce under a prompting of speculation. Therefore, at the very beginning of the launch of stock index futures, it is advisable to require a prudent margin setting in the transaction. With an increase of the transaction volume and market liquidity, investors will gain a better understanding of the risk in spread trading and get a better acquisition with spread estimation. Under such mature market environment, we suggest both EWMA method by selecting a precisely estimated decay factor and ARMA-GARCH model. The former has an obvious advantage of simple application and the latter is distinctive in accuracy and robustness. Once the fixed decay factor under EWMA can not be well estimated for certain reason, ARMA-GARCH model will turn out to be a prior recommendation in the target application. With a further data accumulation and a growing expansion in sample size, EVT can be applied as an effective method for examining the sufficiency of margin level with its prominent advantage in an accurate measuring of extreme risk. However, we have to express a conservative attitude towards a widely application of EVT on a newly open market due to its strict requirement for chronic data accumulation.

## References

1. Login, F.: Optimal Margin Levels in Futures Markets: Extreme Price Movements. Journal of Futures Markets 19, 127–152 (1999)
2. Hill, B.M.: A simple general approach to inference about the tail of a distribution. Annals of Statistics 3, 1163–1174 (1975)
3. Cotter, J.: Margin exceedences for European Stock Index Futures using Extreme Value Theory. Journal of Banking and Finance 25, 1475–1502 (2001)
4. Li, X., Song, X., Pan, X.: Empirical Research on Stock Index Futures Margin Setting of China Based on the Method of Extreme Value Theory. Statistics & Information Forum 04, 42–47 (2006)
5. Fan, Y.: EWMA method for estimating VaR of stock market and its application. Forecasting 3, 34–37 (2001)
6. Liu, Y., Chi, G., Yu, F., Sun, S., Wang, Y.-g.: Forecast model of futures price based on GARCH and EWMA. Journal of Harbin Institute of Technology 9, 1572–1575 (2006)
7. Bollerslev, T.: Generalized Autoregressive Conditional Heteroskedasticity. Journal of Econometrics 31, 307–327 (1986)
8. Engle, R.F.: Autoregressive Conditional Heteroscedasticity with Estimates of the Variance of UK Inflation. Econometrica 50, 987–1008 (1982)
9. Nelson, D.B.: Conditional Heteroskedasticity in asset returns: a new approach. Econometrica 59, 347–370 (1991)

10. Gilli, M., Kellizi, E.: An Application of Extreme Value Theory for Measuring Risk. Computational Economics 27(1), 1–23 (2006)
11. Pickands, J.: Statistical Inference Using Extreme Order Statistics. Annals of Statistic 3, 119–131 (1975)
12. Huisman, R., Koedijk, K., Pownall, R.: VaR-x: fat tails in financial risk management [J]. Journal of Risk 1(1), 47–61 (1998)
13. Kupiec, P.H.: Techniques for verifying the accuracy of risk measurement models [J]. Journal of Derivative 3(2), 73–84 (1995)

# A Study on Multi-word Extraction from Chinese Documents

Wen Zhang[1], Taketoshi Yoshida[1], and Xijin Tang[2]

[1] School of Knowledge Science, Japan Advanced Institute of Science and Technology, 1-1
Ashahidai, Tatsunokuchi, Ishikawa 923-1292, Japan
{zhangwen, yoshida}@jaist.ac.jp
[2] Institute of Systems Science, Academy of Mathematics and Systems Science,
Chinese Academy of Sciences, Beijing 100080, P.R. China
xjtang@amss.ac.cn

**Abstract.** As a sequence of two or more consecutive individual words inherent with contextual semantics of individual words, multi-word attracts much attention from statistical linguistics and of extensive applications in text mining. In this paper, we carried out a series studies on multi-word extraction from Chinese documents. Firstly, we proposed a new statistical method, augmented mutual information (AMI), for words' dependency. Experiment results demonstrate that AMI method can produce a recall on average as 80% and its precision is about 20%-30%. Secondly, we attempt to utilize the variance of occurrence frequencies of individual words in a multi-word candidate to deal with the rare occurrence problem. But experimental results cannot validate the effectiveness of variance. Thirdly, we developed a syntactic method based on lexical regularities of Chinese multi-word to extract the multi-words from Chinese documents. Experimental results demonstrate that this syntactical method can produce a higher precision on average as 0.5521 than AMI method but it cannot produce a comparable recall. Finally, the possible breakthrough on combining statistical methods and syntactical methods is shed light on.

**Keywords:** multi-word extraction, word dependency, mutual information, augmented mutual information, syntactical method.

## 1 Introduction

A word is characterized by the company it keeps [1]. That means not only an individual word but also the context of this word should be laid on great emphasis for further textual processing. This simple and direct motivation drives the researches on multi-word which is anticipated to capture the context information from documents. Although multi-word has no satisfactory formal definition, it can be defined as a sequence of two or more consecutive individual words, which is a semantic unit, including steady collocations (proper nouns, terminologies, etc.) and compound words. Usually, it is made up of a group of individual words and its meaning is either changed to be entirely different from (e.g. collocations) or derived by the straightforward composition of the meanings of its parts (e.g. compound words).

Y. Ishikawa et al. (Eds.): APWeb 2008 Workshops, LNCS 4977, pp. 42–53, 2008.

In fact, there are some overlapping between multi-word, collocation, terminology and similar concepts to describe the unique lexical unit in natural language. For this reason, the definition of multi-word is varied according to different purposes [3, 6, 10, 12], while the fundamental idea behind these concepts is the same, that is, to find lexical term that is more meaningful and descriptive than individual word.

Generally speaking, there are mainly three types of methods developed for multi-word extraction. The first one is the linguistic methods which utilized the structural properties of phrases and sentences to extract the multi-words from document. The second one is the statistical methods based on corpus learning for word pattern discovery from documents. The third one is to combine both the linguistic methods and statistical methods.

As for the linguistic methods, Justeson and Katz analyze the grammatical structure of terminology and propose an algorithm for terminological multi-word identification from English texts [2]. Their method regulates the terminology using a regular expression. It is reported that the method can obtain coverage as 97% and at least 77% precision in noun multi-word identification, and 67% noun multi-words are conformed to the regulation given by them. Similar work can also be found in [3, 4].

Statistical methods mostly employed a series of statistical variables on words' frequency and words' position are proposed to measure the possibility of a word pair to be a multi-word[1]. For instance, Smadja used the relative offset of two words' positions occurring in a corpus to determine whether or not they constitute a multi-word [5]. His basic idea of multi-word is that the offset of two words' positions should be a uniform distribution if these two words can not constitute a multi-word. And he reported that this method is quite successful in terminological extraction with an estimated accuracy as 80%. Similar work can also be referred to [6, 9, 14 and 21].

In the aspect of combining the linguistic knowledge and statistical computation, Chen et al use the co-related text segments existing in a group of documents to identify the multi-word terms from traditional Chinese documents [7]. Chinese stop-list is utilized to split the whole sentence into text segments and the statistical measure derived from term frequency and document frequency is used to weight the text segments to determine whether or not longer segments should be further split into short segments as multi-word terms. Their method is surprisingly successful in multi-word extraction from traditional Chinese documents as they declare that their method can obtain a minimum recall as 76.39% and a minimum precision as 91.05%. However, the performance of their method is determined by the quality of the stop-lists specific for the target texts. Park et al combine the linguistic method and statistical method for domain specific glossary extraction in [8]. The linguistic method is used to produce the candidate items and the statistical method is used for multi-word ranking and selection. Similar work can also be found in [11, 15 and 20].

Our contribution in this paper includes three aspects. Firstly, we proposed AMI to measure the words' dependency with goal to cope with the two problems in MI as unilateral co-occurrence and rare occurrence. Secondly, we investigate the effect of

---

[1] Here we just talk about word pair, i.e. bi-gram, because if a method is validated for word pair, it can be accordingly extended to the case of more than two words.

variance on multi-word extraction. Previous work is invested to use predefined threshold for word frequency to cope with the rare occurrence problem but ours is to make use of the variance of words frequencies in a candidate as an alternative solution. Thirdly, we developed a linguistic method for multi-word extraction using syntactic patterns of multi-words.

## 2   Mutual Information, AMI, Variance and the Syntactic Method

In this section, MI is reviewed in order to present its two deficiencies for dependency measure of word pair. Then AMI is proposed to deal with the deficiencies of MI. Especially, variance is attempted to attack the problem of rare occurrence. And the proposed syntactic method is specified.

### 2.1   Mutual Information

Church and Hanks propose the association ratio for measuring word association based on the information theoretic concept of mutual information [9]. In their method, the MI between word $x$ and $y$ was defined as Eq.1.

$$I(x, y) = \log_2 \frac{P(x, y)}{P(x)P(y)} \tag{1}$$

$P(x)$ is the occurrence probability of term $x$ and $P(y)$ is the occurrence probability of word $y$ in a corpus.

   Sproat and Shih develop a purely statistical method using MI to determine the word boundary in Chinese characters [10]. Their algorithm is very successful for word extraction from Chinese text but the limitation of their method is that it can merely deal with words of length with two characters. Yamamoto and Church combine residual inverse document frequency (RIDF) and MI to conduct the word extraction from Japanese text collection and they report that the substrings with higher RIDF and higher MI are more possible to be a Japanese word [11]. Kita et al compare the competence of MI and cost criteria in multi-word extraction from Japanese and English corpus [12]. Their study demonstrates that mutual information tends to extract task-dependent compound noun phrases, while the method of cost criteria tends to extract predicate phrase patterns. Boxing Chen et al use MI to compute the association score of single word pair to automatically align the bilingual multi-word units from parallel corpora of Chinese and English [13]. Their experimental results demonstrate that the performance of MI was varying with different lexicons because not all the source words have their corresponding target phrase in another language but it provided a basis for constructing a translation lexicon in which the source language and the target language are both multi-word phrases. Jian and Gao propose a method based on MI and context dependency for compound words extraction from very large Chinese Corpus and they report that their method is efficient and robust for Chinese compounds extraction [14]. However, there are too many heuristics involved to determine the context dependency and the parameters are difficult to control to obtain a robust performance.

The primary reason of applying MI for multi-word extraction is that it has the support from both information theory and mathematic proof. If word x and word y are independent form each other, i.e. x and y co-occurred by chance, $P(x,y) = P(x)P(y)$, so $I(x, y) = 0$. By analogy, $I(x, y) > 0$ if x and y are dependent of each other. The higher MI of a word pair, the more genuine is the association between two words.

MI has some inherent deficiencies in measuring association. One is the unilateral co-occurrence problem, that is, it only considers the co-occurrence of two words while ignoring the cases that when one word occur without the occurrence of another. In this aspect, Church and Gale provide an example of using MI to align the corresponding words between French word "chambre", "communes" and English word "house" [15]. The MI between "communes" and "house" is higher than "chambre" and "house" because "communes" co-occurred with "house" with more proportion than "chambre" with "house". But the MI does not consider that more absence is with "communes" than "chambre" when "house" occurred. So it determined the incorrect "communes" as the French correspondence of English word "house". The other is concerned with the rare occurrence problem [16]. As is shown in Eq.1, when we assume that $P(x)$ and $P(y)$ are very small value but $I(x, y)$ can be very large despite of the small value of $P(x, y)$ in this situation. That means the dependency between X and Y is very large despite that X and Y co-occur very small times.

In order to compare with AMI method, the traditional MI method is employed to extract multi-words from a Chinese text collection. Usually, the length of the multi-word candidate is more than two, so we need to determine at which point the multi-word candidate can be split into two parts in order to use the traditional MI formula to score the multi-word candidate. To solve this problem, all the possible partitions are generated to separate a multi-word candidate into two parts, and the one which has the maximum MI score is regarded as the most appropriate partition for this multi-word candidate. Although some practical methods are suggested to extract the multi-word in [12], the method of maximum MI score employed here is different from them, because they are bottom up methods from individual words to multi-words, and our method is a top-down method from multi-word candidate to individual words or other smaller multi-word candidates. However, essentially, they have same back principle, i.e., to split the multi-word candidate into two components, and use MI to rank the possibility of its being a multi-word. For a multi-word candidate as a string sequence $\{x_1, x_2, ..., x_n\}$, the formula for computing its MI score is as follows.

$$MI(x_1, x_2, ..., x_n) = \underset{1 \leq m \leq n}{Max} \{ \log_2 \frac{P(x_1, x_2, ...x_n)}{P(x_1, ..., x_m)P(x_{m+1}, ..., x_n)} \} \tag{2}$$

where $m$ is the breakpoint of multi-word which separates $\{x_1, x_2, ..., x_n\}$ into two meaningful parts, $(x_1, ..., x_m)$ and $(x_{m+1}, ..., x_n)$. Moreover, we can determine whether or not $(x_1, ..., x_m)$ and $(x_{m+1}, ..., x_n)$ are two meaningful words or word combinations by looking up the single word set and multi-word candidate set we established in the previous step. With the maximum likelihood estimation, $P(x_1, x_2, ..., x_n) = F(x_1, x_2, ..., x_n)/N$ ( $N$ is the total word count in the corpus), so the MI method can be rewritten as follows.

$$MI = \log_2 N + \underset{1 \leq m \leq n}{Max} \{\log_2 F(x_1, x_2, ..., x_n) - \log_2 F(x_1, ..., x_m) - \log_2 F(x_{m+1}, ..., x_n)\} \qquad (3)$$

The traditional MI score method for the multi-word candidate ranking in this paper is based on Eq.3.

## 2.2 Augmented Mutual Information

To attack the unilateral co-occurrence problem, AMI is proposed and defined as the ratio of the probability of word pair co-occurrence over the product of the probabilities of occurrence of the two individual words except co-occurrence, i.e., the possibility of being a multi-word over the possibility of not being a multi-word. It has the mathematic formula as described in Eq.4.

$$AMI(x, y) = \log_2 \frac{P(x, y)}{(P(x) - P(x, y))(P(y) - P(x, y))} \qquad (4)$$

AMI has an approximate capability in characterizing the word pair's independence using MI but in the case of word pair's dependence with positive correlation, which means that the word pair is highly possible to be a multi-word, it overcome the unilateral co-occurrence problem and distinguish the dependency from independency more significantly. To attack the problem of rare occurrence problem, we defined the AMI for multi-word candidate more than two words as Eq.5.

$$AMI(x, y, z) = \log_2 \frac{P(x, y, z)}{(P(x) - P(x, y, z)(P(y) - P(x, y, z))(P(z) - P(x, y, z))} \qquad (5)$$

In practical application for a sequence $(x_1, x_2, ..., x_n)$, $P(x_1, x_2, ..., x_n) = p$, $P(x_1) = p_1$, $P(x_2) = p_2$, ..., $P(x_n) = p_n$, we have

$$AMI(x_1, x_2, ..., x_n) = \log_2 \frac{p}{(p_1 - p)(p - p_2)...(p - p_n)} \qquad (6)$$

By maximum likelihood estimation,

$$AMI(x_1, x_2, ..., x_n) = \log_2 \frac{p}{(p_1 - p)(p_2 - p)...(p_n - p)} = \log_2 \frac{F/N}{(F_1 - F)(F_2 - F)...(F_n - F)/N^n}$$

$$= \log_2 \frac{N^{n-1} F}{(F_1 - F)(F_2 - F)...(F_n - F)} = (n-1)\log_2 N + \log_2 \frac{F}{(F_1 - F)(F_2 - F)...(F_n - F)} \qquad (7)$$

$$= (n-1)\log_2 N + \log_2 F - \sum_{i=1}^{n} \log_2 (F_i - F)$$

$F$ is the frequency of $(x_1, x_2, ..., x_n)$ and $F_n$ is the frequency of $x_n$. $N$ is the number of words contained in the corpus, it is usually a large value more than $10^6$. In Eq.7, $\log_2 N$ actually can be regarded as how much the AMI value will be increased by when one more word is added to the sequence. It is unreasonable that $\log_2 N$ is a large value and it makes the AMI is primarily dominated by the length of sequence. In our method,

$\log_2 N$ is replaced by $\alpha$ which is the weight of length in a sequence. Another problem with Eq.7 is that in some special case we have $F_i = F$ and $F_i - F = 0$, these special cases would make the Eq.8 meaningless. For this reason, Eq.7 is revised to Eq.8.

$$AMI(x_1, x_2, ..., x_n) = (n-1)\alpha + \log_2 F - \sum_{i=1}^{m} \log_2 (F_i - F) + (n-m)\beta \tag{8}$$

$m$ is the number of single words whose frequency are not equal to the frequency of the sequence in the corpus. $\beta$ is the weight of the single word whose frequency is equal to the frequency of the sequence. This kind of single word is of great importance for a multi-word because it only occurs in this sequence such as "Lean" to "Prof. J. M. Lean".

## 2.3 Variance

In our method, AMI is a primary measure for ranking the multi-word candidate. Besides AMI, the variance among the occurrence frequencies of the individual words in a sequence is also used to rank the multi-word candidate as the secondary measure. The motivation of adopting variance is that multi-word often occurs as a fixed aggregate of individual words. And the result of this phenomenon is that the variance of frequencies of individual words affiliated to a multi-word would be less than that of a random aggregate of individual words. This makes the variance could be used to attack the rare occurrence problem because the sequence with rare words relative to other frequent words in the same sequence will have a greater variance than those do not have. The variance of a sequence is defined as

$$V(x_1, x_2, ..., x_n) = \frac{1}{n}\sum_{i=1}^{n}(F_i - \overline{F})^2 \quad \overline{F} = \frac{1}{n}\sum_{i=1}^{n}F_i \tag{9}$$

## 2.4 The Syntactic Method

The syntactic method we employed here is made from Justeson and Katz's repetition and noun ending. But it is not the same as their method because Chinese syntactic structure is different from that of English. Feng et.al proposed a method based on the accessor variety of a string to extract the unknown Chinese words from texts [17]. Actually, their method is frequency based because the left accessor variety and the right accessor variety are determined by the number of different characters at the position of head and tail of the string. They used the number of the characters of these two positions to conduct the word segmentation, i.e. delimit word from a string, on Chinese characters so that the consecutive meaningful segments are extracted as words. Their idea is to great extent similar with ours proposed here. But our method does not need an outsourcing dictionary to support adhesive character elimination because adhesive characters have limited influences on multi-word extraction as their short lengths. As the counterpart of repetition in Chinese, any two sentences in a Chinese text were fetched out to match their individual words to extract the same consecutive patterns of them. And we extracted the multi-words from the extracted repetitive patterns by regulating their end words as nouns. The algorithm developed to extract Chinese multi-words is as follows.

**Algorithm 1.** A syntactic method to extract the multi-word from Chinese text

```
Input:
  s₁, the first sentence;
  s₂, the second sentence;
Output:
  Multi-words extracted from s1 and s2;
Procedure:
  s₁ = {w₁,w₂,…,wₙ} s₂ = {w₁',w₂',…,wₘ'} k=0
  for each word wᵢ in s₁
    for each word wⱼ' in s₂
      while(wᵢ = wⱼ')
        k++
      end while
      if k>1
        combine the words from wᵢ to wᵢ₊ₖ' as the same con-
        secutive pattern of s₁ and s₂ as s₃ = {w₁'',w₂'',…,
        wₖ₊₁''}
      End if
    End for
  End for
  p = |s₃|;
  for word wp'' in s3
        if wp'' is a noun
          return {w1'',…,wp''} as the output of this
  procedure;
        else p = p-1;
        end if
        if p is equal to 1
          return null as the output of this procedure;
        end if
  end for
```

# 3   Multi-word Extraction from Chinese Documents

In this section, a series of experiments were conducted with the task to extract the multi-words from Chinese documents to evaluate the proposed methods in Section 2. Basically, we divided the experiments as two groups: one is to evaluate the statistical methods as MI, AMI and Variance; the other is to evaluate the syntactic method.

## 3.1   System Overview of Multi-word Extraction Using MI, AMI and Variance

The multi-word extraction using statistical methods includes primarily three steps. The first step is to generate the multi-word candidate from text using N-gram method. The second step is to rank the multi-word candidates by statistical method, respectively.

The third step is to conduct multi-word selection at different candidate retaining level (clarified in Section 3.4). Figure 1 is the implementation flow chart for multi-word extraction from Chinese documents using MI, AMI and Variance, respectively.

## 3.2 Chinese Text Collection

Based on our previous research on text mining [18, 19], 184 Chinese documents from Xiangshan Science Conference Website (http://www.xssc.ac.cn) are downloaded and used to conduct multi-word extraction. The topics of these documents mainly focus on the basic research in academic filed such as nanoscale science, life science, etc so there are plenty of noun multi-words (terminologies, noun phrases, etc) in them. For all these documents, they have totally 16,281 Chinese sentences in sum. After the morphological analysis[2] (Chinese is character based, not word based), 453,833 individual words are obtained and there are 180,066 noun words.

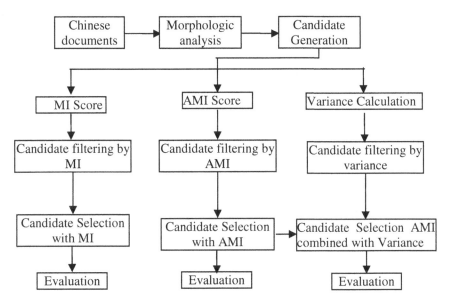

**Fig. 1.** Multi-word extraction from Chinese documents using the statistical methods MI, AMI and Variance, respectively

## 3.3 Candidate Generation

The multi-word candidates are produced by the traditional N-gram method. Assuming we have a sentence after morphological analysis as "A B C DE F G H." and H is found as a noun in this sentence, the candidates will be generated as "G H", "F G H", "E F G H", "D E F G H" and "C D E F G H" because multi-word usually has a length of 2-6 individual words.

---

[2] We conducted the morphological analysis using the ICTCLAS tool. It is a Chinese Lexical Analysis System. Online: http://nlp.org.cn/~zhp/ ICTCLAS/codes.html

**Definition 1. Candidate Set** is a word sequence set whose elements are generated from the same root noun in a sentence using n-gram method.

For example, "G H", "F G H", "E F G H", "D E F G H" and "C D E F G H" construct a candidate set generated from the root noun "H". At most only one candidate from a candidate set can be regarded as the exact multi-word for a root noun.

### 3.4 Requisites for Evaluation

The AMI formula in Eq.8 is used to rank the multi-word candidates. Here, $\alpha$ and $\beta$ was predefined as 3.0 and 0. $\alpha$ is a heuristic value derived from our experiments on computing the AMI of all candidates. $\beta$ is set to 0 as it contributes an unit in length to the candidate so that the AMI value of the sequence with this individual word will be greater than that of the sequence without this individual word. Also the variance of each candidate is calculated out for the secondary measure.

**Definition 2. Candidate Retaining Level (CRL)** regulates at what proportion the multi-word candidates with highest AMI are retained for further selection.

In order to match the multi-word given by our methods and the multi-word given by human experts, approximate matching is utilized.

**Definition 3. Approximate matching.** Assumed that a multi-word is retrieved from a candidate set as $m_1 = \{x_1, x_2, ..., x_p\}$ and another multi-word as $m_2 = \{x_1', x_2', ..., x_p'\}$ was given by human identification, we regard them as the same one if $\dfrac{|m_1 \cap m_2|}{|m_1 \cup m_2|} \geq \dfrac{1}{2}$.

The reason for adopting approximate matching is that there are certainly some trivial differences between the multi-word given by computer and human identification because human has more "knowledge" about the multi-word than computer such as common sense, background context, etc.

### 3.5 Multi-word Extraction Using MI, AMI and Variance

Multi-word extraction using MI employed Eq.3 to rank the candidates and AMI method employed Eq.8 to rank the candidates. It should be noticed that in the Variance method as shown in Fig 1, we used AMI as the first filtering criterion and variance as secondary measure. That is, if a candidate has the greatest AMI and the least variance in its candidate set concurrently, this candidate will be regarded as a multi-word. Otherwise, it will not be regarded as a multi-word and no multi-word comes out from this candidate set.

We varied the CRL for each method at different ratio as 70%, 50% and 30% so that the performances of the above three methods can be observed at dynamic settings. A standard multi-word base for all documents is established. 30 of 184 papers are fetched out randomly from text collection as test samples and the performances of examined methods are observed from them.

Table 1 shows the experimental results from MI, AMI and Variance, respectively. It can be seen that recall is decreasing while precision is increasing when CRL declines from 0.7 to 0.3. The decrease of recall can be convincingly explained because fewer candidates are retained. And the increase on precision clarifies that multi-words actually have higher AMI than the candidates which are not the multi-words. On average, the greatest recall is obtained as 0.8231 at CRL 0.7 with AMI method and the greatest precision is obtained as 0.2930 at CRL 0.3 also with AMI method. This illustrates that AMI outperforms MI and Variance convincingly on all the parameter settings. The performance of MI and Variance are comparable on the whole: the precision of Variance is significantly higher than MI and the Recall of MI is better than MI method.

The motivation of variance is that the individual words of a fixed phrase which is a multi-word usually have a less variance in their occurrence frequencies. We reasonably speculate that the variance would improve the precision in multi-word extraction although the improvement of recall is not ensured. However, the experiment results did not validate our assumption of variance and the fact is that variance would reduce both the recall and the precision in multi-word extraction. We conjecture that multi-words may not have the same properties as that of the fixed phrases, that is, the individual words of a multi-word do not usually have the least variance among its candidate set although they have a low variance. For instance, assuming all individual words of a candidate are rare occurrence words, the variance of that candidate certainly is the least one in its candidate set. This candidate cannot be regarded as a multi-word because of its low occurrence.

**Table 1.** Performances of strategy one and strategy two on multi-word extraction from Chinese text collection at different CRLs. Av is the abbreviation of "average"; R is the abbreviation of "Recall"; P is the abbreviation of "Precision"; F is the abbreviation of "F-measure".

| | MI | | | AMI | | | Variance | | |
|------|--------|--------|--------|--------|--------|--------|--------|--------|--------|
| CRL | Av-R | Av-P | Av-F | Av-R | Av-P | Av-F | Av-R | Av-P | Av-F |
| 0.7 | 0.7871 | 0.2094 | 0.3272 | 0.8231 | 0.2193 | 0.3425 | 0.5621 | 0.2019 | 0.2913 |
| 0.5 | 0.5790 | 0.2174 | 0.3105 | 0.6356 | 0.2497 | 0.3515 | 0.3951 | 0.2317 | 0.2832 |
| 0.3 | 0.2652 | 0.2375 | 0.2419 | 0.3878 | 0.2930 | 0.3229 | 0.2040 | 0.2553 | 0.2160 |

### 3.6   Evaluation for the Syntactic Method

Table 2 is the evaluation results of the proposed syntactic method. It can be seen that the syntactic method can produce a higher precision than any one of the above statistic methods. However, the recall of this syntactical method cannot compete with the statistical methods. Furthermore, we should notice here that the F-measure of the syntactic method is also greater that any one of our statistical methods. That means the syntactic method can produce a more balancing recall and precision than those from statistical methods. In other words, the advantage of statistic method is that it can cover most of the multi-words in Chinese texts despite of its low precision but the syntactic method can produce a highly qualified multi-word extraction although its coverage is limited.

**Table 2.** Performance of syntactic method on multi-word extraction from Chinese text collection. Av is the abbreviation of "average".

| Av-Precision | Av-Recall | Av-F-measure |
|:---:|:---:|:---:|
| 0.3516 | 0.5520 | 0.4152 |

## 4  Concluding Remarks and Future Work

In this paper, we proposed three methods for multi-word extraction as AMI, Variance and a syntactic method. AMI is proposed to attack the two deficiencies inherent in MI. We pointed out that AMI has an approximate capability to characterize the independent word pairs but can amplify the significance of dependent word pairs which are possible to be multi-words. Variance was attempted to attack the problem of rare occurrence because individual words belonging to a multi-word may usually co-occur together and this phenomenon will make the variance of their occurrence frequencies very small. A syntactic method based on the simple idea as repetition and noun ending is also proposed to extract the multi-words from Chinese documents.

Experimental results showed that AMI outperforms both MI and Variance in statistical method. Variance cannot solve the problem of rare occurrence effectively because it can improve nether precision nor recall in multi-word extraction from Chinese documents. The syntactic method can produce a higher precision than the statistic methods proposed in this paper but its recall is lower than the latter. Based on this, we suggest that the performance of extraction could be improved if statistical methods are used for candidate generation and the linguistic method for further multi-word selection.

As far as the future work was concerned, the performance of multi-word extraction is still of our interest, that is, statistical and linguistic methods will be combined according their advantages in multi-word extraction. More experiments will be conducted to validate our hypotheses, especially on the solution of rare occurrence problem. Moreover, we will use the multi-words for text categorization and information retrieval, so that the context knowledge could be integrated into practical intelligent information processing applications.

## Acknowledgments

This work is supported by Ministry of Education, Culture, Sports, Science and Technology of Japan under the "Kanazawa Region, Ishikawa High-Tech Sensing Cluster of Knowledge-Based Cluster Creation Project" and partially supported by the National Natural Science Foundation of China under Grant No.70571078 and 70221001.

## References

1. Firth, J.R.: A Synopsis of Linguistic Theory 1930-1955. Studies in Linguistic Analysis. Philological Society. Blackwell, Oxford (1957)
2. Justeson, J.S., Katz, S.M.: Technical terminology: some linguistic properties and an algorithm for identification in text. Natural Language Engineering 1(1), 9–27 (1995)

3. Bourigault, D.: Surface Grammatical Analysis for the Extraction of Terminological Noun Phrases. In: Proceedings of the 14th International Conference on Computational Linguistics, Nantes, France, pp. 977–981 (1992)
4. Kupiec, J.: MURAX: A robust linguistic approach for question answering using an on-line encyclopedia. In: Proceedings of the Sixteenth Annual International ACM Conference on Research and Development in Information Retrieval, Pittsburgh, PA, USA, June 27 - July 1, 1993, pp. 181–190 (1993)
5. Smadja, F.: Retrieving collocations from text: Xtract. Computational Linguistics 19(1), 143–177 (1993)
6. Church, K.W., Robert, L.M.: Introduction to special issue on computational linguistics using large corpora. Computational Linguistics 19(1), 1–24 (1993)
7. Chen, J.S., et al.: Identifying multi-word terms by text-segments. In: Proceedings of the seventh international conference on Web-Age information Management Workshops (WAIMV 2006), HongKong, pp. 10–19 (2006)
8. Park, Y.J., et al.: Automatic Glossary Extraction: Beyond Terminology Identification. In: Proceedings of the 19th international conference on Computational linguistics, Taiwan, pp. 1–17 (2002)
9. Church, K.W., Hanks, P.: Word association norms, mutual information, and lexicography. Computational Linguistics 16(1), 22–29 (1990)
10. Sproat, R., Shinh, C.: A statistic method for finding word boundaries in Chinese text. Computer Processing of Chinese and Oriental Language 4(4), 336–351 (1990)
11. Yamamoto, M., Church, K.W.: Using suffix arrays to compute term frequency and document frequency for all substrings in a corpus. Computational Linguistics 27(1), 1–30 (2001)
12. Kita, K., et al.: A comparative study of automatic extraction of collocations from Corpora: mutual information vs. cost criteria. Journal of Natural Language Processing 1(1), 21–29 (1992)
13. Chen, B.X., Du, L.M.: Preparatory work on automatic extraction of bilingual multi-word units from parallel corpora. Computational Linguistics and Chinese Language Processing 8(2), 77–92 (2003)
14. Zhang, J., et al.: Extraction of Chinese Compound words: An experiment study on a very large corpus. In: Proceedings of the second Chinese Language Processing Workshop, HongKong, pp. 132–139 (2000)
15. Church, K.W., William, A.G.: Concordances for parallel text. In: Proceedings of the seventh Annual Conference of the UW Center for the New OED and Text research, Oxford, pp. 40–62 (1991)
16. Christopher, D.M., Hinrich, S.: Foundations of Statistical natural language processing, pp. 178–183. MIT Press, Cambridge (2001)
17. Feng, H.D., et al.: Accessor Variety Criteria for Chinese Word Extraction. Computational Linguistics 30(1), 75–93 (2004)
18. Zhang, W., Tang, X.J., Yoshida, T.: Web text mining on A Scientific Forum. International Journal of Knowledge and System Sciences 3(4), 51–59 (2006)
19. Zhang, W., Tang, X.J., Yoshida, T.: Text classification toward a Scientific Forum. Journal of Systems Science and Systems Engineering 16(3), 356–369 (2007)
20. Daille, B., et al.: Towards Automatic Extraction of Monolingual and Bilingual Terminology. In: Proceedings of the International Conference on Computational Linguistics, Kyoto, Japan, August 1994, pp. 93–98 (1994)
21. Fahmi, I.: C Value Method for Multi-word Term Extraction. Seminar in Statistics and Methodology. Alfa-informatica, RuG, May 23 (2005),
   http://odur.let.rug.nl/fahmi/talks/statistics-c-value.pdf

# Extracting Information from Semi-structured Web Documents: A Framework

Nasrullah Memon[1,3], Abdul Rasool Qureshi[2], David Hicks[1], and Nicholas Harkiolakis[3]

[1] Department of Computer Science and Engineering
Aalborg University
Niels Bohrs Vej 8, DK-6700 Esbjerg, Denmark
[2] Royal Cyber Inc., Karachi, Pakistan
[3] Hellenic American University Athens, Greece

**Abstract.** This article aims to automate the extraction of information from semi-structured web documents by minimizing the amount of hand coding. Extraction of information from the WWW can be used to structure the huge amount of data buried in web documents, so that data mining techniques can be applied. To achieve this target, automated extraction should be utilized to the extent possible since it must keep pace with a dynamic and chaotic Web on which analysis can be carried out using investigative data mining or social network analysis techniques. To achieve that goal a proposed framework called Spiner will be presented and analyzed in this paper.

**Keywords:** Counterterrorism, dark web, iMiner, investigative data mining, spiner.

## 1 Introduction

The bulk of today's extraction systems identify a fixed set of Web resources and then rely on hand coded wrappers [1] to access the resource and parse its response. The information extraction system should be dynamic enough to identify and extract the information from even unfamiliar web sources in order to scale with the growth of the Internet. In this article we propose a framework for extracting information from semi-structured Web documents.

Spiner is the proposed framework/system that is presented in this paper and can minimize the amount of hand coding required for the identification and extraction of information from different web resources. It doesn't eliminate the need for website wrappers needed for parsing but favors XML based wrappers against a site. The Spiner subsystem "XML Application Engine", as shown in Figure 1, provides user graphical facilities to prepare XML wrappers for corresponding sites and their application to extract information. Designing a website wrapper can be as simple as selecting the HTML and designing a pipeline structure (see next section for details) with a graphical user interface. Spiner's web robot uses web indices stored in a database to search and provide the XML Application Engine with the data input. Spiner can use any search engine(s) to populate the web indices database. Spiner's learning engine

Y. Ishikawa et al. (Eds.): APWeb 2008 Workshops, LNCS 4977, pp. 54–64, 2008.

identifies the different patterns in those XML wrappers to extend its learning of parsing the websites. The Spiner Learning Engine keeps on learning HTML data structures, to minimize the writing of XML Wrappers. The XML application engine consults the Spiner learning engine, to help the one who is writing the XML wrappers for the site.

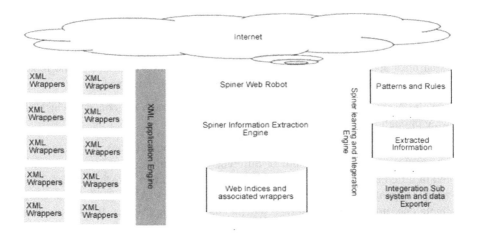

**Fig. 1.** System Architecture of Spiner

When Spiner's learning engine encounters anything, for example, a combination of user text and tags, it prompts the user with all of the results found against that pattern in the patterns database, and the user only needs to select suitable ones from it or write a new one. In case a user designs a new parsing schema against that pattern, the result is saved in the patterns and rules database to be used later against that combination. In case of the absence of user input against any pattern in the database, the learning engine identifies the best one from the database, if it has one, and continues further. In this way Spiner's learning subsystem takes care of parsing for something which has not been wrapped up by any of the wrappers yet. This type of learning experience can be used by the system to parse the new resources whose wrappers are not yet prepared.

Spiner's integration subsystem takes care of its integration with various investigative software applications, so that the extracted information can be used for analysis and investigation. The extracted information is transformed into XML or CSV (comma separated values) files, and exported to other systems like iMiner [2] or NetMiner [3], so that the interesting and useful results can be deducted.

## 2  Pipeline Structure

The pipeline structure is actually something, which standardizes the structuring of HTML. To understand how it accomplishes the task let us first understand its three main constituents.

## 2.1 Components of the Pipeline Structure

*Pipeline:* The pipeline is analogous to pipelines which are used for water supplies in daily life. The only difference is that in Spiner HTML characters will flow in it. As in normal Life every pipeline ends up at some reservoir, where the contents that are flowing in it can be saved. The pipeline has valves to support the distribution of the contents it carries to different reservoirs based on different rules. Any combination of HTML tags and text can be the rule on which reservoirs are changed.

*Valve:* A valve is a type of switch which is used to join two Pipelines. These switches have an open and closed condition over which transmission is shifted to the joined pipeline and back to originating pipeline on which that switch is fixed.

*Reservoir:* A reservoir is a storage place where we can store the contents which are flowing through pipelines, i.e. HTML.

When the HTML flows through the Pipeline, various combinations of tags and text results in the opening and closure of different valves, information is separated and stored in different storage reservoirs. The pipeline structure as shown in figure 2 can be used to separate any links which any html page contains. Although this example is a simple one, it demonstrates how a pipeline data structure works. In the upcoming section a relatively more complex example is also described.

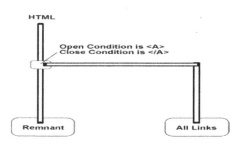

**Fig. 2.** Pipeline Data structure dividing html page in two parts links and remaining html

In this example, HTML is flown through a pipeline which has only one valve which opens up every time a <A> tag flows through pipeline and closes on encountering a </A> tag. This main pipeline ends up at the storage reservoir Remnant. The valve connects the main pipeline with another pipeline which ends up at the storage reservoir all links. Actually whenever the valve is opened the transmission is shifted from the main pipeline to the pipeline which ends at the all links reservoir and whenever the valve is closed the main pipeline carries the html to remnant. We can even separate links to any level by adding valves in the pipeline which carries links. In this way we can identify meaningful html data, separate out and then rejoin different things stored in different reservoirs to any structure. The valve condition can be as complex as is needed and thousands of valves can be fixed on one pipeline.

## 2.2  Why a Pipeline Structure?

To standardize parsing of HTML documents we needed a data structure in which any type of parser can be specified. SAX parsers are there to parse the HTML but can't be used for standardizing extraction of the information from thousands of sources. Much coding is also needed to write SAX based wrappers for thousands of websites which urge the need of a high level framework which is built on the SAX or the DOM Parser to extract information. With a pipeline data structure we can specify any type of rule in XML with an open and close condition which is composed of a combination of HTML components. These valves can be used to identify and separate out values of fields and records lying there in HTML text and the values of these fields and records will be united to form the Structured Information from any website. For joining these field values and records are mapped with XML wrappers which actually wrap up targeted structure.

We can call these wrappers representing the targeted structure as Entity Wrappers because they wrap entities.

Coding can be reduced to a minimum level by programming a GUI which forms an XML Wrapper, which is then used by Spiner's information extraction engine to extract the information.

It is noted that the CSS is used for formatting HTML documents. Rapidly changing sites usually keep changing style sheets to change their layout and GUI. Whenever a site's layout or GUI is changed, it is actually the CSS where the change takes place. The ID's, class names and tag names are the same which are used in the CSS to connect it to HTML components. If the rules for the opening and closure of valves are based on CSS ID's, class names and tag names like we have done in the threats site example, Spiner's pipeline structure can adjust to site changes automatically and we don't have to write a new wrapper for the updated site.

## 3  Designing XML Wrappers with Spiner

Spiner's application engine's GUI is formulated to assist XML wrapper design. The GUI is designed in such a way that it only prompts the user to identify the targeted information and then calculates the unique combination of HTML components at the back end. The CSS styles precedence, so that changes in web layouts can be automated within the Spiner application engine. The engine back-end will be persisting those rules by transforming them into XML automatically. These rules will be used to prepare valves in the pipeline data structure. The GUI can also show the user dynamically the pipeline structure, like we have done in Figure 2 and 4, using any open source visualization framework. This will help the user to comprehend what s/he has accomplished so far and what is still needed.

## 4  Dark Web and Counterterrorism Model

Spiner can be a useful system to utilize in a dark web analysis scenario. The system architecture to accommodate dark web analysis is shown in Figure 3. The wrappers are designed for each of the different terrorist group websites, terrorism databases and government information sites. Spiner web robots will spider those sites and then structured information can be extracted from HTML data provided by the Spiner web robot. That information is used by the integrated investigation system iMiner [2] to deduce analyst-friendly results. For example we can identify a hidden hierarchy using iMiner to uncover which dark sites influence other dark sites in a dark web analysis. The simulation of terrorism databases is also shown in the next section. So Spiner can play a vital role in dark web analysis.

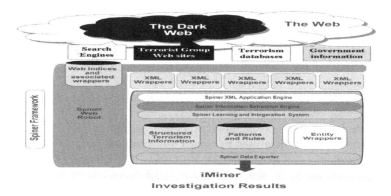

**Fig. 3.** Spiner Model to carry out Dark Web analysis

## 5  Example

The wrapper described in Figure 4 can be used to effectively parse down the HTML at http://www.trackingthethreats.com to extract any entities, relations, metadata and other available info.

Spiner will automatically separate all of them in different reservoirs, from which they can be exported to iMiner for analyzing, visualizing and destabilizing the terrorist networks [5] [6].

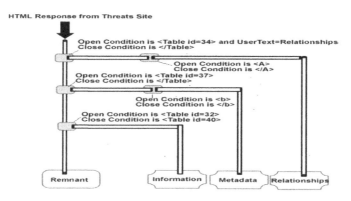

**Fig. 4.** Implementing the above pipeline site

Spiner's Export Subsystem will identify and consolidate, normalize and transform the information such that it is readable and importable to iMiner. iMiner's import facilities will then be used to import that data for analysis.

## 6  iMiner Knowledge Base

The focus of the knowledge base we have developed is the agglomeration of publicly available data and integration of the knowledge base with an investigative data mining software prototype. The main objective is to investigate and analyze terrorist networks to find hidden relations and groups, prune datasets to locate regions of interest, find key players, characterize the structure, trace a point of vulnerability, detect the efficiency of a network and to discover the hidden hierarchy of non-hierarchical networks.

The iMiner knowledge base (see Figure 5) consists of various types of entities. Here is an incomplete list of the different entity types:

- Terrorist organizations such as Al Qaeda
- Terrorists such as Osama Bin Ladin, Ramzi Yousef, etc.
- Terrorist facilities such as Darunta Training Camp, Khalden Training Camp, etc.
- Terrorist events/ attacks such as 9/11, WTC terrorist attack 2003, etc.

The dataset also contains various types of relations connecting instances of different entity types. Here is a partial list of the various relation types:

- MemberOf: instances of terrorist can be affiliated with various instances of terrorist organization.
- FacilityOwner: instances of terrorist facility are usually run by instances of terrorist organizations.
- FacilityMember: instances of terrorist are linked to various instances of terrorist facilities if the terrorist instance attended/ spent some time at the facility.
- ClaimResponsibility: instances of terrorist organization are linked to the instances of terror attacks they claim responsibility for.
- ParticipatedIn: instances of terrorist that may have participated in instances of terror attacks.

The iMiner system applies a spider management system as well as spiders to import data that is available on the Web. We have developed a prototype system to get information from online repositories and save it in its knowledge base for analysis.

The analysis of relational data is a rapidly growing area within the larger research community interested in machine learning, knowledge discovery, and data mining. Several workshops [7, 8, 9] have focused on this precise topic, and another DARPA research program — Evidence Extraction and Link Discovery (EELD) —focuses on extracting, representing, reasoning with, and learning from relational data.

We stored data in the knowledge base in the form of triples,

*<subject, object, relationship>*

where subject and object may be entities of interest and relationship is a link between the two entities. These triples are also used in the Resource Description Framework (RDF).

The binary–relational view regards the universe as consisting of entities with bi-nary–relationships between them. An entity is anything which is of interest and can be identified. A binary–relationship is an association between two entities. The first entity in a relationship is called the subject and the second entity is called the object. A relationship is described by identifying the subject, the type of relationship, and object for example: Bin Laden is leader of Al Qaeda can be written as (Bin Laden. leaderOf .Al Qaeda).

N–ary relationships such as Samad got bomb making training from Zarqawi may be reduced to a set of binary–relationships by the explicit naming of the implied en-tity. For example:

training # 1. trainee .Samad
training #1. trainer .Zarqawi
training #1. got training of .bomb making

There are several advantages and disadvantages of using binary relational storage [10] such as:

- The simple interface between a binary relational storage structure and other modules of a database management system consists of three procedures:
  o insert (triple); (ii) delete (partial specification of triple); (iii) retrieve (partial specification of triple)
- Retrieval requests such as "list of all terrorists that met at Malaysia before 9/11" are met by issuing a simple retrieval call (terrorist. met at Malaysia be-fore 9/11. ?) which delivers only that data which has been requested.

Multi-attribute retrieval may be inefficient. The use of a binary relational structure for multi-attribute retrieval might not be the most efficient method (irrespective of the way in which the structure is implemented). To illustrate this, consider the example: "retrieve (the names of) all young (25 years old) married terrorists involved in the 9/11 attacks who attended flight schools and got terrorist training in Afghanistan". Using a binary-relational structure, one could issue the retrievals:

(? .marital status. Married)
(? .age. 25)
    (? .involved in. 9/11 terrorist attacks)
    (? .attended. flight schools)
    (? .got terrorist training. Afghanistan)

and merge the resultant sets to obtain the required data. This is probably much faster than having to do a sequential search through the whole database, which would be necessary if the database were held as a file of records (name, age, marital status, etc.) ordered on name. The sociogram of the Al Qaeda network integrated in the software prototype iMiner is depicted in Figure 5.

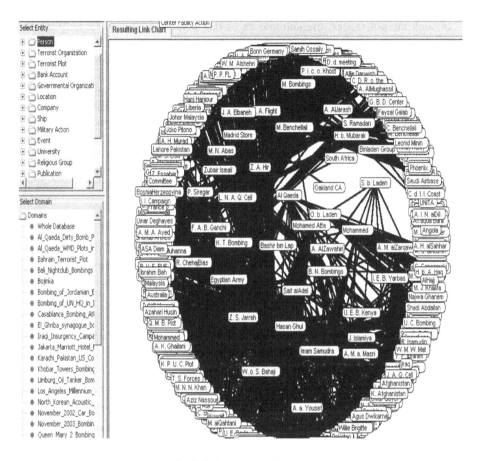

**Fig. 5.** iMiner Knowledge base

We have harvested terrorists' information from the Internet. The iMiner system is bundled with a powerful spidering engine to update its knowledge base by harvesting the information present on different sources on the Web. At present it only supports one data source, as discussed above, but it can be expanded to support many data sources as per requirement.

## 7  Automating a Dynamic Price Engine

E-commerce has emerged as the most important field of commerce. In a world of people where a high percentage of them use the internet regularly, the importance of e-commerce cannot be ignored. The advent of B2B sites has provided end users with a kind of bargaining on the internet, where they can get a glimpse of the prices at which different vendors and distributors are selling. On one end these sites have facilitated end-users, while on the other end the cost is paid by the distributors and vendors by regularly maintaining their shopping carts, fine-tuning their prices and profits to keep their self in the market.

In such circumstances, the automation of this price tuning is absolutely necessary. The automation can be a rule based engine which on the basis of predefined rules, tunes the prices of the products which are on catalogue. The database of rules can be either maintained with RDBMS technology or XML. The other main variable involved in the processing of such an engine is the prices with which the competitors in the market are selling the same product. The lowest prices registered on B2B price comparison sites are entities of premiere interest too for such rule based engines. The engine can then perform optimization of prices and profits based on business constraints specified in the shape of rules, the pricing information gathered from the internet, and the original prices of the product as shown in Figure 6.

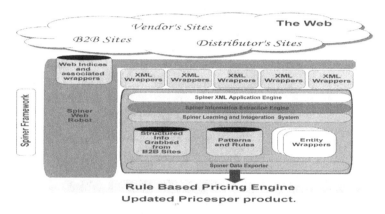

**Fig. 6.** Spiner's Model for automating a dynamic pricing engine to equip e-commerce systems

The ever changing nature of the internet sites has made automation of product information and price extraction very difficult, if not impossible. The layouts of sites keep on changing, which adds a lot to the maintenance cost of an Extraction Engine and development never ends for such a system. Spiner can be viewed as an affordable solution to this business problem. The GUI of Spiner's Application Engine makes it possible for the user to express parsing logic in XML wrappers and thus eliminating the need for coding. Designing the XML Wrapper is as easy as selecting the targeted information and the Application Engine calculates the combination of tags to uniquely identify the information as its own. As the Application Engine is described in proceeding sections, it will not be redefined here.

The other problem for such a system is how to identify a targeted product in the absence of any standard. There are many standards being designed originally to address this problem like MFG Number, MPN, UPC, and SKU. The problem with those standards is illustrated in the following business scenario.

Suppose "*A's shopping*" is an ecommerce site for distributor "*A*" which uses an MPN number as a Product Identification standard. "*B's shopping*" is an ecommerce system for A's competitor which uses an SKU for product identification. In such a case, matching A's products with B's products can be very dangerous. It is seen in the current e-commerce market that each distributor doesn't differentiate the standards the same, so even if *B* starts using the MPN standard, SKU's can be stored as MPN numbers and thus make it a dangerous choice to be used in matching. The distributors are also seen adding a suffix to MPN numbers and polluting it enough to make it difficult to be used for matching. To illustrate this situation, we may consider a product such as the Toshiba Satellite A135-S4527 Notebook.

| Product | Price comparison site | Product Identifier |
|---|---|---|
| Toshiba Satellite A135-S4527 Notebook | www.pricegrabber.com | MPN: PSAD0U05400P |
| Toshiba Satellite A135-S4527 Notebook | www.shopping.com | MPN: A135-S4527 UPC: 032017833005 |

Hence we cannot accomplish the task by merely matching the MPN standards. The shopping site also provides the UPC number but due to the fact that the price-grabber site doesn't follow the UPC standard we cannot use it as well. These types of difficulties can be countered by using detailed specification matching. Spiner can be used to parse down their detailed specification, transform them into name value pairs and then matching the attributes of each product to identify the same product.

# 8   Conclusion

In this article we have proposed a new framework for extracting information from semi-structured web pages. The proposed framework (Spiner) can be a useful system to utilize in a dark web analysis scenario. The system architecture to accommodate a dark web analysis was described and analyzed. Efficient wrappers were designed for different terrorist group websites, terrorism databases and government information sites. The system's web robots crawl those sites and extract structured information from HTML data provided by the system's web robots. Also, an alternative use of the proposed system architecture for e-commerce purposes has been described and alyzed.

From the analysis it is evident that Spiner will be capable of handling the dynamic and chaotic character of the present day World Wide Web by using investigative data mining enhanced with social network analysis techniques. A future research direction that we intend to follow is the implementation of Spiner into a fully function and customizable tool that will be further used by academics and businesses. The practical aspects of the framework will be explored and its suitability for other areas of research will be investigated.

# References

1. Krulwich, B.: The bargainfinder agent: Comparison price shopping on the internet. In: Williams, J. (ed.) Bots and other Internet Beasties. SAMS.NET (1996)
2. Nasrullah, M.: Investigative data mining: Mathematical models for analyzing, visualizing and destabilizing terrorist networks. PhD dissertation, Aalborg University Denmark (2007)

3. Cyram NetMiner: Social Network Analysis Software
   `http://www.netminer.com/NetMiner/home_01.jsp`
4. XML Wrappers: White Paper,
   `http://www.zylab.es/downloads/whitepapers/White%20Paper%20-`
   `%20XML%20Wrappers%20for%20Email%20Archiving.pdf`
5. Memon, N., Hicks, D.L., Larsen, H.L., Rajput, A.Q.K.: How Investigative Data Mining Can Help Intelligence Agencies in Understanding and Splitting Terrorist Networks. In: The Second Workshop on Data Mining Case Studies and Practice Prize:13th ACM SIGKDD International Conference on Knowledge Discovery and Data Mining (KDD 2007), San Jose, August 12-15, 2007, pp. 26–45 (2007)
6. Memon, N., Hicks, D., Larsen, H.L.: Practical Algorithms and Mathematical Models for Analysing, Visualizing and Destabilizing Terrorist Networks. In: 2007 Miltary Communications Conference, Orlando, Florida, October 29-31. IEEE, Los Alamitos (2007)
7. Dzeroski, S., De Raedt, L., Wrobel, S. (eds.): Papers of the Workshop on Multi-Relational Data Mining. The Eighth ACM SIGKDD International Conference on Knowledge Discovery and Data Mining. ACM Press, New York (2002)
8. Getoor, L., Jensen, D. (eds): Learning Statistical Models from Relational Data. In: Proceedings of AAAI 2000 Workshop. AAAI Press, Menlo Park (2000)
9. Jensen, D., Goldberg, H.: Artificial Intelligence and Link Analysis. In: Proceedings of 1998 AAAI Fall Symposium. AAAI Press, Menlo Park (1998)
10. Frost, R.A.: Binary Relational Storage Structures. Computer Journal (1982)

# Discovering Interesting Classification Rules with Particle Swarm Algorithm

Yi Jiang[1,2], Ling Wang[3], and Li Chen[2]

[1] The School of Computer Science, Wuhan University, Wuhan, 430072, China
yijiang78@163.com
[2] The School of Computer Sci. and Tech., Wuhan University of Science and Technology,
Wuhan, 430065, China
[3] Wuhan University of Science and Technology City College, Wuhan, 430083, China

**Abstract.** Data mining deals with the problem of discovering novel and interesting knowledge from large amount of data. It is the core problem in building a fuzzy classification system to extract an optimal group of fuzzy classification rules from fuzzy data set. To efficiently mine the classification rule from databases, a novel classification rule mining algorithm based on particle swarm optimization (PSO) was proposed. The experimental results show that the proposed algorithm achieved higher predictive accuracy and much smaller rule list than other classification algorithm.

**Keywords:** Data mining, Classification, Particle swarm algorithm.

## 1 Introduction

In the last years information collection has become easier, but the effort required to retrieve relevant pieces of it has become significantly greater, especially in large-scale databases. As a consequence, there has been a growing interest in powerful tools capable of facilitating the discovering of interesting and useful information within such a huge amount of data.

One of the possible approaches to this problem is by means of data mining or knowledge discovery from databases (KDD)[1]. Through data mining, interesting knowledge can be extracted and the discovered knowledge can be applied in the corresponding field to increase the working efficiency and to improve the quality of decision making. Classification rule mining is one of the important problems in the emerging field of data mining which is aimed at finding a small set of rules from the training data set with predetermined targets[2]. There are different classification algorithms used to extract relevant relationship in the data as decision trees that operate a successive partitioning of cases until all subsets belong to a single class. However, this operating way is impracticable except for trivial data sets. In literature, several machine learning methods have been applied in solving fuzzy classification rule generation problem. In [3], a method called fuzzy decision trees is presented to extract fuzzy classification rules from data set containing membership values. Since fuzzy decision tree is easily trapped into local optimal, evolutionary algorithms with ability

Y. Ishikawa et al. (Eds.): APWeb 2008 Workshops, LNCS 4977, pp. 65–73, 2008.

of global searching is introduced in the domain [4]. In recent years, evolutionary algorithms(such as genetic algorithm, immune algorithm and ant colony algorithm) have emerged as promising techniques to discover useful and interesting knowledge from database[5]. Rule extraction method based on evolutionary method is widely adopted in recent publications and a survey can be found in [6].

The classification problem becomes very hard when the number of possible different combinations of parameters is so high that algorithms based on exhaustive searches of the parameter space become computationally infeasible rapidly. Especially, there are numerous attempts to apply genetic algorithms(GAs) in data mining to accomplish classification tasks. In addition, the particle swarm optimization (PSO) algorithm[7][8], which has emerged recently as a new metaheuristic derived from nature, has attracted many researchers' interests. The algorithm has been successfully applied to several minimization optimization problems and neural network training. Nevertheless, the use of the algorithm for mining classification rule in the context of data mining is still a research area where few people have tried to explore. The algorithm, which is based on a metaphor of social interaction, searches a space by adjusting the trajectories of individual vectors, called "particles " as they are conceptualized as moving points in multidimensional space. The individual particles are drawn stochastically toward the position of their own previous best performance and the best previous performance of their neighbors. The main advantages of the PSO algorithm are summarized as: simple concept, easy implementation, robustness to control parameters, and computational efficiency when compared with mathematical algorithm and other heuristic optimization techniques. The original PSO has been applied to a learning problem of neural networks and function optimization problems, and efficiency of the method has been confirmed. In this paper, the objective is to investigate the capability of the PSO algorithm to discover classification rule with higher predictive accuracy and a much smaller rule list. In the following sections, a simple review of PSO algorithm will be made, then the algorithm for generation of classification rules is described in details, and at last an experiment is conducted to show the performance of the algorithm.

## 2   PSO Algorithm and Its Binary

### 2.1   PSO Algorithm

PSO is a population-based optimization technique proposed firstly for the above unconstrained minimization problem. In a PSO system, multiple candidate solutions coexist and collaborate simultaneously. Each solution called a particle flies in the problem search space looking for the optimal position to land. A particle, as time passes through its quest, adjusts its position according to its own experience as well as the experience of neighboring particles. Tracking and memorizing the best position encountered build particle's experience. For that reason, PSO possesses a memory (i.e. every particle remembers the best position it reached during the past). PSO system combines local search method (through self experience) with global search methods (through neighboring experience), attempting to balance exploration and exploitation.

A particle status on the search space is characterized by two factors: its position and velocity, which are updated by following equations.

$$V_i[t+1] = \omega V_i[t] + c_1 \times rand(\bullet) \times (P_i - X_i) + c_2 \times Rand(\bullet) \times (P_g - X_i) \tag{1}$$

$$X_i[t+1] = X_i[t] + V_i[t+1] \tag{2}$$

where, $V_i = [v_{i,1}, v_{i,2}, \cdots, v_{i,n}]$ called the velocity for particle i, which represents the distance to be traveled by this particle from its current position; $X_i = [x_{i,1}, x_{i,2}, \cdots, x_{i,n}]$ represents the position of particle i; $P_i$ represents the best previous position of particle i (i.e. local-best position or its experience); $P_g$ represents the best position among all particles in the population $X = [X_1, X_2, \cdots X_N]$ (i.e. global-best position); Rand() and rand() are two independently uniformly distributed random variables with range [0,1]; $c_1$ and $c_2$ are positive constant parameters called acceleration coefficients which control the maximum step size; $\omega$ is called the inertia weight that controls the impact of previous velocity of particle on its current one. In the standard PSO, Equation(1) is used to calculate the new velocity according to its previous velocity and to the distance of its current position from both its own best historical position and its neighbors' best position. Generally, the value of each component in $V_i$ can be clamped to the range $[-v_{max}, v_{max}]$ to control excessive roaming of particles outside the search space. Then the particle flies toward a new position according to(2). This process is repeated until a user-defined stopping criterion is reached.

## 2.2  The Binary Version of PSO Algorithm

As mentioned above, PSO is extended to binary version to cope with discrete optimization problem. The difference between the binary version and original one mainly lies in the particle's encoding style. In the binary version, the ith particle's value of each dimension, denoted by xid, is either 0 or 1. The exact value of xid is determined by a probability which depends on the changing rate vid. The concrete mechanism can be depicted as the following equations:

$$v_{id} = \omega \cdot v_{id} + c_1 \cdot Rand() \cdot (p_{id} - x_{id}) + c_2 \cdot Rand() \cdot (p_{gd} - x_{id}) \tag{3}$$

$$if \ (rand() < S(v_{id})) \ then \ x_{id} = 1 \ else \ x_{id} = 0 \tag{4}$$

where, rand() and Rand() are quasirandom numbers selected from a uniform distribution in [0.0 , 1.0] [9], S( . ) denotes the famous sigmoid function. Through analyzing the above equations, it can be found that the value of $x_{id}$ has a probability of $S(v_{id})$ to be one, and $1 - S(v_{id})$ to be zero.

## 3 The PSO System

The aim is the implementation of a genetic system able to automatically extract an intelligible classification rule for each class in a database, given the values of some attributes, called predicting attributes. Each rule is constituted by a logical combination of these attributes. This combination determines a class description which is used to construct the classification rule.

In this paper, PSO is used to generate fuzzy rules in the form :

IF $(A_1$ is $T^1{}_m)$ and ...and $(A_i$ is $T^i{}_j)$ and ...and $(A_k$ is $T^k{}_n)$ THEN $(C$ is $C_p)$

where $A_i (i = 1, 2...k)$ is called condition attribute, C is conclusion attribute, $T^i{}_j$ is jth fuzzy linguistic value of the condition attribute $A_i$, $C_p$ is the pth fuzzy linguistic value of conclusion attribute C. Our fuzzy PSO algorithm is to be described in this section, and there are mainly the following topics to be discussed in detail for our algorithm.

### 3.1 Particle Encoding

A certain fuzzy classification rule, i.e., a sequence of linguistic terms, can be denoted by pure binary string, namely: if a linguist term is valid, represented with binary value 1 otherwise [4]. At the same time, the binary version PSO in which each particle is encoded into a binary string is adopted in fuzzy PSO, so each particle in the swarm represents a candidate rule. A rule encodingexample for the Saturday Morning Problem [3], If outlook is rain then plan isW_lifting, can be found in fig.1. In fuzzy PSO, the whole particle swarm is divided into several sub-populations according to the number of the conclusion attribute items. In each subpopulation of rules, the particles have the same corresponding component of conclusion attribute, and the initial particles come from two sources: one comes directly from the samples data set, and the other from randomly generating. The purpose of initializing the particle swarm in sucha way is to get a better start point of searching than purely randomly doing it.

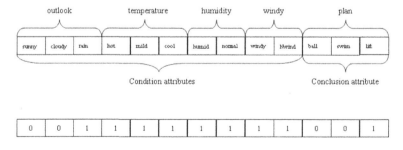

**Fig. 1.** Rule encoding example

## 3.2  Fitness Function

The population is constituted by the possible class descriptions, that is to say sets of conditions on attributes of the objects to classify. In these cases it is possible to use statistical fitness functions [10]. Let us introduce some definitions to formalize the fitness function $f_c$ which assigns a numerical value indicating its correctness to any description d in the description space D. To each description d in D it corresponds a subset of the training set S denoted with $\sigma_D(S)$, i.e. the set of points where the conditions of a rule are satisfied. The correctness depends on the size of $\sigma_D(S)$, covered by the description, the size of the class C representing the points where the prediction of the rule is true and the size of their overlapping region $\sigma_D(S) \cap C$.

Moreover, if we denote with $\sigma_D(S)$ the set of the points in the database in which the conditions are not satisfied and C' the set of the points in which the prediction of the rule is false, the simplest fitness function can be devised as follow

$$f_c = \mid \sigma_D(S) \cap C \mid + \mid \sigma_{D'}(S) \cap C' \mid - (\mid \sigma_{D'}(S) \cap C \mid + \mid \sigma_D(S) + C' \mid) \quad (5)$$

This function can be explained as the difference between the actual number of examples for which the rule classifies properly the belonging or not belonging to the class and the number of examples for which there is an incorrect classification whether the conditions or the prediction are not satisfied.

Besides these statistical factors, the fitness function $f_c$ also contains a further term which takes into account in some way the simplicity of the description. This term is conceived taking in mind Occam's razor[14] "the simpler a description, the more likely it is that it describes some really existing relationships in the database". This concept is incorporated in the function $f_s$ and it is related to the number of nodes ($N_{nodes}$) and the tree depth (depth) of the encoded rules. Namely, this topological term is:

$$f_s = N_{nodes} + depth \quad (6)$$

The total fitness function F considered is:

$$F = (S\_tr - f_c) + \alpha f_s \quad (7)$$

where S_tr is the number of samples in the training set and α varies in [0.0, 1.0]. The choice of its value affects the form of the rules: the closer this value to 1the lower the depth tree and the number of nodes.

## 3.3  Particle Swarm Evolution Process

The process of evolution in our algorithm including the following steps:

(1) Generate at random an initial population of rules representing potential solutions to the classification problem for the class at hand;
(2) Evaluate each rule on the training set by means of a fitness function;

(3) Updating both each individual particle's best position and the best position in the subswarm.

(4) Reinsert these offspring to create the new population;

(5) Repeat steps (3) to (6) until an acceptable classification rule is found or the specified maximum number of generations has been reached;

(6) Repeat steps (2) to (6) until one rule is determined for each class in the database;

(7) Assign each example in the training and in the test sets to one and only one class.

In most of PSO, the whole swarm maintains the global best particle position which is as the common experience of the swarm and influences each particle's trajectory in searching space. As mentioned above, in fuzzy PSO the whole swarm is divided into several sub-swarms, and the corresponding conclusion parts of the particles have the same form. In each sub-swarm, the sub-best position which serves as the common experience of the sub-swarm, is maintained to affect the trajectory of each particle in it. So the formula (3) and (4) should be re-expressed as the following equations:

$$v^j_{id} = \omega \bullet v^j_{id} + c_1 \bullet Rand() \bullet (p^j_{id} - x^j_{id}) + c_2 \bullet Rand() \bullet (p^j_{gd} - x^j_{id}) \qquad (8)$$

$$if \ (rand() < S(v^j_{id})) \ then \ x^j_{id} = 1 \ else \ x^j_{id} = 0 \qquad (9)$$

To take advantage the "memory" ability of PSO, fuzzy PSO extracts rules according to the particles' best individual positions instead of the current particle population, so it is expected to get better rule in fewer iteration times. The process of rule extraction which is similar with that presented in [13] is adopted in our algorithm, and it is depicted in fig.2, where $\tau$, $\delta$ and $\gamma$ are all constants preset by user. When a correct rule is extracted, the cases covered by it are removed out of sample set and the coverage of each rule in the candidate rule group is recomputed. If the sample set is empty, the algorithm is terminated, otherwise extraction process is continued until the candidate rule group is empty. If the latter occurs, fuzzy PSO will initial a new swarm of particles to find correct rule until the termination condition is satisfied. To improve the efficiency of the algorithm, the scale of the particle swarm is self-adaptive to the number of cases in the sample set. Once the particle swarm is initialized, the scale, i.e., the number of particles in the whole swarm, is computed as :

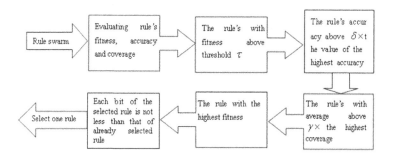

**Fig. 2.** Rule extraction process

$$scale = p_{min} + curr\_num \times (p_{max} - p_{min}) / orig\_num \qquad (10)$$

where, scale is the current number of particles in the swarm, $p_{min}$ and $p_{max}$ denote the minimum and maximum particle numbers in the swarm respectively, curr_num and orig_num denote the current and original numbers of cases in sample set.

## 4 Experimental Results

To thoroughly investigate the performance of the proposed PSO algorithm, we have conducted experiment with it on a number of datasets taken from the UCI repository[9]. In Table 1, the selected data sets are summarized in terms of the number of instances, and the number of the classes of the data set. These data sets have been widely used in other comparative studies.

Table 1. Dataset Used in the Experiment

| Data Set | Instances | Classes |
| --- | --- | --- |
| Ljubljana Breast Cancer | 282 | 2 |
| Wisconsin Breast Cancer | 683 | 2 |
| Tic-Tac-Toe | 958 | 2 |
| Dermatology | 366 | 6 |
| Hepatitis | 155 | 2 |
| Cleveland Heart Disease | 303 | 5 |

We have evaluated the performance of PSO by comparing it with Ant-Miner[15], OCEC(a well-known genetic classifier algorithm)[16]. The first experiment was carried out to compare predictive accuracy of discovered rule lists by well known tenfold cross-validation procedure. Each data set is divided into ten partitions, each method is run ten times, using a different partition as test set and the other nine partitions as the training set each time. The predictive accuracies of the ten runs are averaged as the predictive accuracy of the discovered rule list. Table 2 shows the results comparing the predictive accuracies of PSO, Ant-Miner and OCEC, where the symbol "±" denotes the standard deviation of the corresponding predictive accuracy. It can be seen that predictive accuracies of PSO is higher than those of Ant-Miner and OCEC.

Table 2. Predictive Accuracy Comparison

| Data Set | PSO(%) | Ant-Miner(%) | OCEC(%) |
| --- | --- | --- | --- |
| Ljubljana Breast Cancer | 78.56±0.24 | 75.28±2.24 | 76.89±0.18 |
| Wisconsin Breast Cancer | 98.36±0.28 | 96.04±0.93 | 95.42±0.02 |
| Tic-Tac-Toe | 98.89±0.13 | 73.04±2.53 | 92.51±0.15 |
| Dermatology | 98.24±0.26 | 94.29±1.20 | 93.24±0.12 |
| Hepatitis | 95.75±0.31 | 90.00±3.11 | 91.64±0.23 |
| Cleveland Heart Disease | 79.46±0.34 | 57.48±1.78 | 76.75±0.16 |

In addition, We compared the simplicity of the discovered rule list by the number of discovered rules. The results comparing the simplicity of the rule lists discovered by PSO, Ant-Miner and OCEC are shown in Table 3. As shown in those tables, taking into number of rules discovered, PSO mined rule lists much simpler(smaller) than the rule lists mined by Ant-Miner and OCEC.

**Table 3.** Number of Rules Discovered Comparison

| Data Set | PSO | Ant-Miner | OCEC |
|---|---|---|---|
| Ljubljana Breast Cancer | 6.05±0.21 | 7.10±0.31 | 16.65±0.21 |
| Wisconsin Breast Cancer | 4.23±0.13 | 6.20±0.25 | 15.50±0.13 |
| Tic-Tac-Toe | 6.45±0.37 | 8.50±0.62 | 12.23±0.25 |
| Dermatology | 6.39±0.24 | 7.30±0.47 | 13.73±0.18 |
| Hepatitis | 3.01±0.26 | 3.40±0.16 | 10.73±0.35 |
| Cleveland Heart Disease | 7.15±0.23 | 9.50±0.71 | 15.37±0.42 |

At last, we also compared the running time of PSO with Ant-Miner and OCEC. The experimental results are reported Table 4,as expected, we can see that PSO's running time is fewer than Ant-Miner's and OCEC's in all data sets.The main reason is that PSO algorithm is conceptually very simple and requires only primitive mathematical operators codes. In addition, PSO can be implemented in a few lines of computer codes, those reduced PSO's running time. Insummary, PSO algorithm needs to tune very few algorithm parameters, taking into account both the predictive accuracy and rule list simplicity criteria, the proposed PSO-based classification rule mining algorithm has shown promising results.

**Table 4.** Running Time Comparison

| Data Set | PSO | Ant-Miner | OCEC |
|---|---|---|---|
| Ljubljana Breast Cancer | 31.25 | 55.28 | 46.37 |
| Wisconsin Breast Cancer | 42.35 | 58.74 | 45.25 |
| Tic-Tac-Toe | 38.65 | 61.18 | 52.38 |
| Dermatology | 27.37 | 49.56 | 37.23 |
| Hepatitis | 38.86 | 56.57 | 42.89 |
| Cleveland Heart Disease | 31.83 | 48.73 | 35.26 |

## 5  Conclusions

Classification rule mining is one of the most important tasks in data mining community because the data being generated and stored in databases are already enormous and continues to grow very fast. In this paper, a PSO-based algorithm for classification rule mining is presented. Compared with the Ant-Miner and OCEC in public domain data sets, the experimental results show that the proposed algorithm achieved higher predictive accuracy and much smaller rule list than Ant-Miner and OCEC.

**Acknowledgments.** This research was supported by the National Natural  Science Foundation of China under Grant NO. 60705012.

# References

1. Fayyad, U.M., Piatetsky-Shapiro, G., Smyth, P.: From Data Mining to Knowledge Discovery: an Overview. In: Advances in Knowledge Discovery & Data Mining, pp. 1–34. MIT Press, Cambridge (1996)
2. Quinlan, J.R.: Induction of Decision Trees. Machine Learning 1, 81–106 (1986)
3. Yuan, Y., Shaw, M.J.: Induction of fuzzy decision trees. Fuzzy Sets and Systems 69, 125–139 (1995)
4. Yuan, Y., Zhuang, H.: A generic algorithm for generating classification rules. Fuzzy Sets and Systems 84, 1–19 (1996)
5. Freitas, A.A.: Data Mining and Knowledge Discovery with Evolutionary Algorithms. Springer, Berlin (2002)
6. Freitas, A.A.: A survey of evolutionary algorithms for data mining and knowledge discovery. In: Advances in Evolutionary Computation. Springer, Heidelberg (2001)
7. Eberhart, R.C., Kennedy, J.: A New Optimizer using Particle Swarm Theory. In: Proc. 6th Symp. Micro Machine and Human Science, Nagoya, Japan, vol. 43, pp. 39–43 (1995)
8. Clerc, M., Kennedy, J.: The particle swarm—Explosion, stability, and convergence in a multidimensional complex space. IEEE Trans. on Evolutionary Computation 6(1), 58–73 (2002)
9. Kennedy, J., Eberhart, R.: A discrete version of the particle swarm algorithm. In: IEEE Conference on Systems, Man, and Cybernetics, pp. 4104–4109 (1997)
10. Holsheimer, M., Siebes, A.: Data mining: the search for knowledge in databases, Technical Report, CS-R9406, CWI, Amsterdam, The Netherlands (1994)
11. Kaya, M.: Multi-objective genetic algorithm based approaches for mining optimized fuzzy association rules. Soft Computing 7(10), 578–586 (2005)
12. Derkse, W.: On simplicity and Elegance, Eburon, Delft (1993)
13. Ma, X., et al.: The path planning of mobile manipulator with genetic-fuzzy controller in flexible manufacturing cell. In: Proceedings of the 1999 IEEE International Conference on Robotics & Automation, pp. 329–334 (1999)
14. Hettich, S., Bay, S.D.: The UCI KDD Archive (1999), http://kdd.ics.uci.edu
15. Parpinelli, R.S., Lopes, H.S., Freitas, A.A.: Data Mining with an Ant Colony Optimization Algorithm. IEEE Transactions on Evolutionary Computing 6, 321–332 (2002)
16. Liu, J., Zhong, W.-C., Liu, F., Jiao, L.-C.: Classification Based on Organizational Coevolutionary Algorithm. Chinese Journal of Computers 26, 446–453 (2003)
17. Zhao, X., Zeng, J., Gao, Y., Yang, Y.: Particle Swarm Algorithm for Classification Rules Generation. In: Sixth International Conference on Intelligent Systems Design and Applications 2006. ISDA 2006, October 2006, vol. 2, pp. 957–962 (2006)
18. Misra, B.B., Biswal, B.N., Dash, P.K., Panda, G.: Simplified Polynomial Neural Network for classification task in data mining. In: Evolutionary Computation, 2007. CEC 2007. IEEE Congress, September 25-28, 2007, pp. 721–728 (2007)

# Improving the Use, Analysis and Integration of Patient Health Data

David Hansen, Mohan Karunanithi, Michael Lawley, Anthony Maeder,
Simon McBride, Gary Morgan, Chaoyi Pang, Olivier Salvado, and Antti Sarela

The Australian e-Health Research Centre, CSIRO ICT Centre, Brisbane, Australia
{Firstname.Surname}@csiro.au

**Abstract.** Health Information Technologies (HIT) are being deployed world-wide to improve access to individual patient information. Primarily this is through the development of electronic health records (EHR) and electronic medical records (EMR). While the proper collection of this data has reached a high level of maturity, the use and analysis of it is only in its infancy. This data contains information which can potentially improve treatment for the individual patient and for the cohort of patients suffering a similar disease. The data can also provide valuable information for broader research purposes. In this paper we discuss the research contributions we are making in improving the use and analysis of patient data. Our projects include the analysis of physiological data, the extraction of information from multi-modal data types, the linking of data stored in heterogeneous data sources and the semantic integration of data. Through these projects we are providing new ways of using health data to improve health care delivery and provide support for medical research.

**Keywords:** ambulatory monitoring, cohort analysis, data linking, electronic health records, semantic integration, time series analysis.

## 1 Introduction

Today there is a large amount of data being collected systemically about a person's health. Each time a person interacts with the health service (which occurs increasingly for a chronic disease rather than an acute health event), data is recorded. This data can take the form of doctors' notes, vital signs and images acquired from machines, coded data entered in a database, survey data from the patient, or reports from experts about a medical examination.

In this paper we discuss some of the novel applications being built at The Australian E-Health Research Centre (AEHRC)[1], which extract more information from a patient's aggregate health data. All of these projects aim to provide information to the right person at the right time and in the right form for effective health care use.

---

[1] http://aehrc.com/ last accessed 16 April 2008.

Y. Ishikawa et al. (Eds.): APWeb 2008 Workshops, LNCS 4977, pp. 74–84, 2008.

## 2   Physiological Data

Patient physiological data are being monitored in an increasing number of health care situations. While this data can provide significant information about the health status of a patient, using and analyzing it can be a time consuming task that is currently done either sporadically or by the application of fairly simple computer analysis methods.

### 2.1   Ambulatory Monitoring of Cardiovascular Patients

With the increasing number of patients living with chronic diseases and multiple co-morbidities, the health system needs to develop clinical solutions for managing these patients in efficient ways. Cardiovascular diseases are one such set of chronic diseases where the integration of clinical factors and monitoring technology has the ability to reduce burden on health care systems and improve the quality of service [10].

The scope of the Care Assessment Platform (CAP) project is to develop data transformation and analysis processes which can be used in a sustainable home–based care situation. This platform will initially provide software solutions for cardiovascular disease management and prevention based on data obtained from ambulatory monitoring devices. The primary benefits anticipated from this model are the assistance rendered to patients and carers, and a reduction in the rate of hospitalisation.

**Current Trial.** Currently AEHRC is finalising a clinical trial in conjunction with the Northside Health Service District[2] in Brisbane, Australia, across the local community primary healthcare setting. Preliminary results [2] indicate that by using ambulatory monitoring devices measuring physical activity and ECG on cardiac rehabilitation patients, it is possible to automatically extract clinically relevant information on a patient's behaviour and physical status. This information can be used for:

- Discriminating walking from other high intensity activities,
- Calculating walking speed to determine index of functional capacity,
- Calculating energy expenditure due to walking,
- Quantifying the amount and duration of walking events.

Other developed measures can be produced based on this information:

- Patient functional capacity including gait patterns,
- Detection and measurement of sit–to–stand transition speed,
- Various measures for heart rate variability (HRV).

All these measures indicate different aspects of patient's physiological status during rehabilitation and can be used to support decisions and actions of the team of care professionals, as well as the patients and their families for self management.

**Planning New Trials.** Following the current project which is monitoring patients in hospital while they recover, a new trial is being planned to monitor the patients in their home environment. It is planned that the new home–based cardiac care model will utilise and integrate novel information technology software, algorithms, home

---

[2] http://www.health.qld.gov.au/wwwprofiles/northside.asp last accessed 16 April 2008.

care devices and systems. The project will test the home–based model by planning and conducting a large scale randomised controlled clinical study on four different home and hospital–based cardiac care models with and without the use of monitoring technology. In addition to innovations in process development and evidence on clinical and economic outcomes, the trial is expected to create new clinical knowledge on physiological signal patterns, exercise, diet, lifestyle, and behaviour of cardiovascular patients in a home care setting.

## 2.2  Analysis of Physiological Signals under Anesthetic

Another area of clinical activity where a patient's physiological signals are monitored is during anaesthesia [6]. AEHRC is working with the Department of Anaesthetics at the Royal Brisbane and Womens' Hospital (RBWH)[3] to improve patient safety by providing analyses of physiological time series data, acquired from monitoring of patients in the operating theatre while they are under anesthetic.

Currently coarse-grained automated alarms together with clinical observation are used to monitor patients who are under critical care regimes such as in the operating theatre or intensive care wards. The nature of the alarms means that they sound either very often, and are hence devalued as a clinical aid, or they occur later than would give the anesthetist optimum time to act. Meanwhile, clinical observation of the signals also could lead to missing subtle pattern changes which would again not give the anesthetist the best time to act.

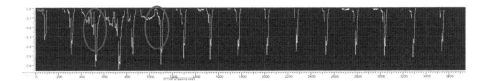

**Fig. 1.** Analysis of ECG data for changes in patterns

To help clinicians make prompt and proper decisions from the time series data, we propose a novel and intelligent data analysis system that makes best use of the physiological signals by efficiently compressing the time series data and is able to detect and compare interesting patterns. For real-time analysis there exist two major challenges: fast creation of concise yet effective representation of the time series data; and online support of complex time series queries such as novelty detection, motif recognition and trend forecast. Aiming at these two objectives, we have proposed a linear compression algorithm which employs the wavelet technique to generate a concise synopsis of the original data, importantly where that synopsis guarantees the data is within a certain error [9]. Currently our experimental results demonstrate that our data analysis system is efficient and promising for clinical use. It also suggests opportunity for more complex time series analysis based upon wavelet synopses.

---

[3] http://www.health.qld.gov.au/rbwh/ last accessed 16 April 2008.

# 3 Multi-modal Data

A large proportion of the data captured by the health system is as a set of readings off a machine, such as the physiological data in the last section, which can be treated by direct numerical analysis. However, there may also be data of highly structured and more complex data types, typically captured via human intervention. This might include images or free text reports, which require a clinical expert to interpret. The projects described in this section are designed to help those clinicians with the analysis of this complex data.

## 3.1 Free-Text Data: Analysis of Pathology Reports

The analysis of images or pathology samples by experts results in free text reports being sent to the patient's clinician. The Cancer Stage Interpretation System (CSIS) project focuses on analyzing free text pathology reports to obtain a preliminary cancer stage. This will support systems for cancer management, both for individual patients and population-level analyses.

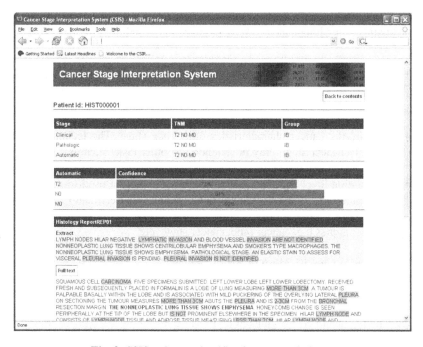

**Fig. 2.** CSIS software classifies free text pathology

The cancer "stage" is a categorisation of its progression in the body, in terms of the extent of the primary tumour and any spreading to local or distant body sites. While staging has a fundamental role in cancer management, due to the expertise and time required and the multi–disciplinary nature of the task, cancer patients are not always routinely staged. By automating the collation, analysis, summarisation and classification of relevant patient data, the reliance on expert clinical staff can be lessened, improving the efficiency and availability of cancer staging.

The CSIS project, in collaboration with the Queensland Cancer Control Analysis Team (QCCAT)[4], has produced a software prototype system for automatic pathological staging of lung cancer and this has been developed on a set of 710 lung cancer patients. The system inputs one or more free text reports for a patient describing surgical resections of the lung, and outputs a pathological T and N stage. In addition, an extract is produced consisting of sentences that were found to contribute to the final staging decision, and their relationship to criteria from the formal staging guidelines for lung cancer. The system has been formally trialled in a clinical setting on a previously unseen set of 179 lung cancer cases. The trial validated the automatic stage decisions with the stages assigned by two expert pathologists [7]. The results obtained in the trial have motivated the development of a production-quality system suitable for deployment within cancer registries. Currently this work is being extended to classify M-stage from radiology reports. The work is also examining ways of improving on the classification by making use of mappings from the free text to a clinical ontology, in this case SNOMED CT[5].

Our initial work has investigated the classification of text content in patient reports to assist with staging lung cancer. Longer term research will investigate extending this in three ways:

- Extensions to handle other data and cancer types. Initial work is focusing on staging lung cancer using text reports for radiology and histology. Opportunities exist to extend this to bowel and other cancers, and also to use information extracted from other forms of data, e.g. medical image contents.
- Classifying cancer characteristics other than stage. The techniques used to classify cancer stage may be extended to other tasks such as filtering of patient data, e.g. diagnosis of cancer, or classification of cancer types.
- Population-level analyses. Statistical models may be used to identify trends and anomalies in cancer patient demographics or treatment / response characteristics, based on metadata extracted through the automatic content analysis techniques e.g. cancer type, cancer stage, etc.

## 3.2   Image Data: Classification of Brain Images

Recent advances in imaging technologies may help to diagnose brain disorders earlier. Delaying onset of major dementia by 5 years could reduce new cases by 50% with cumulative health cost savings of up to $13.5 billion by 2020 in Australia[6].

---

[4] http://www.health.qld.gov.au/cancercontrol/docs/qcc_strat05-10.pdf last accessed 16 April 2008.
[5] The Systematized Nomenclature of Medicine-Clinical Terms (SNOMED CT) standard is managed by the International Health Terminology Standards Development Organisation (IHTSDO): see http://www.ihtsdo.org/our-standards/snomed-ct/ last accessed 16 April 2008.
[6] http://www.alzheimers.org.au/upload/EstimatesProjectionsNational.pdf last accessed 16 April 2008.

The causes of dementia are not well understood and current diagnosis is difficult because there are as yet no known biological markers. The relatively advanced loss of cognitive function necessary for current clinical diagnosis of dementia generally results in irreversible neuronal dysfunction. If objective evidence of AD pathological lesions could be found early (before there is evidence of cognitive or behavioural symptoms), appropriate treatment and care could be provided, resulting in delayed onset or prevention of AD.

AEHRC's biomedical imaging team is developing key technologies for in vivo quantitative assessment of Amyloid-A (Aß) deposition suspected to be an early marker of AD. We are collaborating through CSIRO's Preventative Health Flagship with AIBL[7], the Australian Imaging, Biomarker and Lifestyle cluster study. AIBL has enrolled more than 1000 volunteers for psycho-cognition and blood biomarker evaluation and includes 200 volunteers who will have brain-image scans using Pittsburgh compound B (11C-PIB), a novel Positron Emission Tomography (PET) biomarker. Other information from different imaging modalities will also be acquired from those volunteers such as fluorodeoxyglucose (18FDG) PET imaging, and Magnetic Resonance Imaging (MRI).

Specifically, the biomedical imaging team is developing a library of image processing algorithms that can be called from our core software MilxView, a generic medical imaging viewer. From MilxView individual or large batch of image analysis tasks can be scheduled and run automatically. An example of one such algorithm is an improved method for estimating cortical thickness in brain images [14].

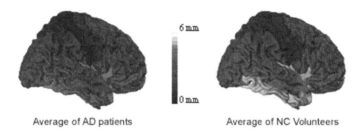

Average of AD patients                    Average of NC Volunteers

**Fig. 3.** Brain images for visualizing neuro-degeneration

Our vision is to provide clinicians with an easy-to-use and fully automatic software tool. The physicians would read the scans from a patient, and we would compute several quantitative measurements from the images otherwise hard or impossible for a human to obtain. From our study on the AIBL patients we will be able to benchmark each individual patient against the typical age-matched individual and provide to physicians relevant statistics. This kind of information should be valuable not only to help in treatment planning of individual patients, but also to help design therapies and test scientific hypothesis.

---

[7] http://www.aibl.nnf.com.au/page/home last accessed 16 April 2008.

## 4  Data Integration

Data is being captured within the health system at an increasing rate. However, it is often captured in disparate systems and is difficult to access. Hence, a number of initiatives are examining how to build information systems which are able to provide an integrated view of the data about a patient while preserving access constraints [3,8]. Integration can happen at many levels, and there are a large number of technical and knowledge management issues which must be addressed to achieve true integration. However, new enabling technologies can provide mechanisms to provide an integrated view of the data within a reasonable time and for relatively little effort.

### 4.1  Data Linking with HDI

A major problem with distributed data sets is that they are held by more than one custodian and that the patient identifiers used are different. Here we discuss a product for health data integration and linking (HDI) [4] which has been developed to support cross-organisational clinical research. HDI offers the ability to integrate health information across organisational boundaries without needing to surrender that information to a centralised database/warehouse. These features provide support to clinicians and data custodians who have legal, ethical and organisational obligations to the health information they control.

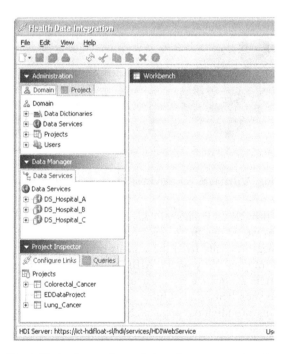

**Fig. 4.** HDI software gives role based access to patient data

Most health organisations maintain investments in health information resources (e.g. databases) that provide data for clinical treatment, research, audit, etc. However, as organisations increase the amount of collaborative research they conduct, the ability to share these information resources across organisational boundaries becomes important.

The HDI Remote Domain feature set enables collaborating organisations to share health information in a manner which supports the legal, ethical and organisational obligations of the data custodians. HDI ensures the security of data as it travels across Internet Protocol networks by using Public Key Infrastructure (PKI) technologies to encrypt network connections. The collaboration protocol implemented by HDI ensures that no data is shared without having express consent of the organization's IT Department and the data custodian, that all access to data is in the context of an approved activity and that each organisation maintains control over its data at all times. HDI has been deployed within a number of organisations who use the product to support collaborative research. In one example, data sets from an ambulance service and a hospital have been linked and integrated to provide a clinical research team with the ability to study an integrated data set. In this and other cases, the whole data set has proved far more valuable than the sum of its parts.

## 4.2  Providing for Semantic Data Integration

While HDI is able to provide a linked data set, with some mapping of data to a common terminology, this will not provide a semantically enabled view of the data. This will only come through the use of a standard but complex terminology which allows the capture data at the appropriate clinical level, and that also encodes the relationships between different clinical concepts [12].

SNOMED CT (developed originally by the American College of Pathologists) has been adopted as the standard for describing health data, in the expectation that collecting data using a single, large ontology will increase the level of understandability and completeness of that data. However, even such a large ontology does not cover all possible concepts which need to be dealt with. Previous studies have shown [13] that mapping existing data to SNOMED CT is possible although with incomplete coverage, and so many current efforts are focused on building extensions for clinical concepts and areas which are not already covered. Once the data is available in a consistent manner for search and retrieval, the actual interface presented to end-users must encourage use.

Since the HDI tool provides a linked data set, this raises the question of how we can provide a semantic layer on data which is already captured. SNOMED CT provides a mechanism to add a set of new concepts to the ontology. These sets of concepts and relationships are known as extensions. While many extensions are built from existing terminologies or from scratch, extensions built from existing data sets may provide a faster way to more complete coverage of the concepts used to capture health data. We have built a mapping tool which allows an extension to be made by mapping data to SNOMED CT concepts. The tool provides the data sources available for building an extension in one part, a search box for retrieving SNOMED codes on the right hand side and a "mappings" box in the middle. An individual data item is then equated to a single concept or an expression editor allows post-coordinated expression to be built. The mapping tool builds an extension based on the mapping and adds it to the existing terminology. The extended terminology then needs to be classified, to check for consistency and provide a full set of relationships. Since most classifiers either cannot work on a large ontology such as SNOMED CT or are too slow, we have implemented our own classification software. Snorocket [5] is a new implementation of the Dresden algorithm [1] which is able to classify SNOMED CT in around 60 seconds, as opposed to over 30 minutes with the fastest existing implementations. Figure 5 shows SNOMED CT with an extension for World Health Organisation International Classification of Diseases (ICD)[8] codes.

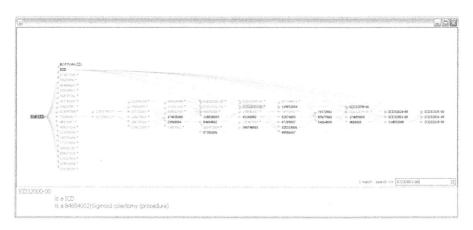

**Fig. 5.** Example of a subset of SNOMED CT with extension

Once multiple data sets can be queried using a single ontology, it becomes possible to ask more complex queries that take into account the semantics between concepts. Example queries include: "find me all patient events which have a code of 73761001 or one of its children". This will find events which map to a colonoscopy (73761001) or one of its children (for example a sigmoidopscopy, 24420007). This we map to a "surveillance event" in the patient timeline.

### 4.3  Novel Applications Using Semantically Integrated Data

Improving the human-computer interface is one of the largest challenges for HIT today [11]. An intuitive interface based on the intrinsic nature of the overall data relationships and allowing different levels of depth in accessing the data would be desirable.

A simple example of a patient data access system using the above concepts is now presented. In our example, we use synthetic data about patients receiving clinical treatment for the diagnosis and treatment of colorectal cancer (CRC). We use the HDI software described above to access the distributed data from six different databases and our mapping tool to make extensions based on the existing data sets. We then use the relationships encoded in the extended SNOMED CT ontology in querying the

---

[8] http://www.who.int/classifications/icd/en/ last accessed 16 April 2008.

data. These queries take the form of finding events in the patient data which are linked to diagnosis and treatment of CRC, such as for example the query for the colonoscopy events given above.

One way of viewing this progression is through a timeline of "health events" which display data from multiple databases against a timeline of when the health event which the data describes occurred. We have used the semantic layer provided by the SNOMED CT mappings to understand if an event relates to surveillance, diagnostic or primary or secondary treatment. The health professional is able to view the patient timeline for a single patient and drill down to retrieve data from original data sources.

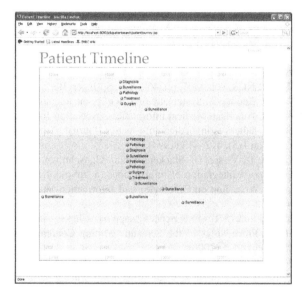

**Fig. 6.** Patient timelines for events in diagnosis and treatment of Colorectal Cancer

## 5 Conclusion

In this paper we have discussed some of the novel software tools being built by The Australian e-Health Research Centre, with the aim of turning the growing data deluge into information and knowledge, and so contribute to improved treatment for patients.

## References

1. Baader, F., Lutz, C., Suntisrivaraporn, B.: CEL - a polynomial-time reasoner for life science ontologies. In: Furbach, U., Shankar, N. (eds.) IJCAR 2006. LNCS (LNAI), vol. 4130, pp. 287–291. Springer, Heidelberg (2006)
2. Boyle, J., Bidargaddi, N., Sarela, A., Karunanithi, M.: Ambulatory care delivery for chronic disease management. In: Proceedings of the 6th IASTED International Conference on Biomedical Engineering, Innsbruck, Austria, pp. 384–389 (2008)

3. Churches, T., Christen, P.: Some methods for blindfolded record linkage. BMC Medical Informatics and Decision Making 4(9), 17 (2004)
4. Hansen, D.P., Pang, C., Maeder, A.: HDI: integrating health data and tools. Journal of Soft Computing 11(4), 361–367 (2007)
5. Lawley, M.: Exploiting fast classification of SNOMED CT for query and integration of health data. In: Proceedings of KR-MED 2008, Phoenix, USA (to appear, 2008)
6. Nunez, C.: Advanced techniques for anesthesia data analysis. Seminars in Anesthesia, Perioperative Medicine and Pain 23(2), 121–124 (2004)
7. McCowan, I.A., Moore, D.C., Nguyen, A.N., Bowman, R.V., Clarke, B.E., Duhig, E.E., Fry, M.J.: Collection of cancer stage data by classifying free-text medical reports. Journal of the American Medical Informatics Association 14(6), 736–745 (2007)
8. O'Keefe, C.M., Yung, M., Baxter, R.: Privacy-preserving linkage and data extraction protocol. In: Proceedings of the 2004 ACM Workshop on Privacy in the Electronic Society, Washington, DC, USA, pp. 94–102 (2004)
9. Pang, C., Zhang, Q., Hansen, D., Maeder, A.: Building data synopses within a known maximum error bound. In: Dong, G., Lin, X., Wang, W., Yang, Y., Yu, J.X. (eds.) APWeb/WAIM 2007. LNCS, vol. 4505, pp. 463–470. Springer, Berlin (2007)
10. Pare, G., Jaana, M., Sicotte, C.: Systematic review of home telemonitoring for chronic diseases. Journal of the American Medical Informatics Association 14(3), 269–277 (2007)
11. Sittig, D.F.: Grand challenges in medical informatics. Journal of the American Medical Informatics Association 1(5), 412–413 (1994)
12. Wache, H., Vogele, T., Visser, U., Stuckenschmidt, H., Schuster, G., Neumann, H., Hubner, S.: Ontology-based integration of information - a survey of existing approaches. In: Proceedings of the Workshop on Ontologies and Information Sharing, Seattle, USA, pp. 108–117 (2001)
13. Wade, G., Rosenbloom, S.T.: Experiences mapping a legacy interface terminology to SNOMED CT. In: Proceedings of the Semantic Mining Conference on SNOMED CT, Copenhagen, Denmark, p. 5 (2006)
14. Zuluaga, M., Acosta, O., Bourgeat, P., Salvado, O., Hernandez, M., Ourselin, S.: Cortical thickness measurement from magnetic resonance images using partial volume estimation. In: Reinhardt, J.M., Pluim, J.P.W. (eds.) Proceedings of SPIE: Image Processing, vol. 6914, 69140J1–69140J8 (to appear, 2008)

# DM-Based Medical Solution and Application

Junjie Liang and Yucai Feng

Wuhan Dameng Database Corp. Ltd., Wuhan Hubei 430074, China
ljj@dameng.com

**Abstract.** Wuhan DAMENG Database Limited Corporation specializes in developing, producing, selling and serving DBMS, with products successful in many application areas. He has proposed many advanced solutions for medical information system and achieved many applications.

**Keywords:** DM Database, Medical Solution, Digital Hospital.

The DAMENG Database Limited Corporation specializes in developing, selling and serving DBMS for 20 years with initial China Certification of Double-Soft Enterprise. He has been supported by many national ministries, commissions and province city halls, such as National Development and Reform Commission, National Ministry of Science and Technology, Ministry of Information Industry, General Armament Department and Ministry of Public Security.

The products of DAMENG have succeeded in many application areas, such as Military, Public Security, Fire Fighting, Tax Administration, Finance, National Territory, Electronic Governmental Affairs, Manufacturing, Electronic Commerce, Education and Medical Health.

On the basis of market requirement, DAMENG will be engaged in developing data warehouse, data exchange, data mining and data management, with the goal of manufacturing Chinese database elaboration and constructing information security platform.

Advanced medical system not only symbolizes a nation development, but also shows citizen life grade. In China, medical information development around "Gold Health" project has been brought into operation. On medical information development being carried forward, many medical systems with Chinese characteristics have been realized, including health service system, medical security system, medical monitor system, etc.

The national basis software DM DBMS has behaved many benefits in medical information development, which will contribute to standardizing medical system as soon as possible. In medical information development, DM can play many important roles with good features of DM DBMS having creative technology with the core technique, supporting high security up to B1, high compatibility, high reliability, high efficiency and usable tools. Furthermore, the DM Corporation has a professional and quality service team for all the customers with low cost.

DM Corporation has proposed many advanced solutions for medical information system, with software development experience and advanced management conception

Y. Ishikawa et al. (Eds.): APWeb 2008 Workshops, LNCS 4977, pp. 85–86, 2008.

for many years. On the basis of DM database, we can construct different management system for different requirement. The system can be HIS, CIS, PACS, LIS and ICU from different business such as out-patient service, stay in hospital, medical image check, etc. For community health service system, DM- based solutions can support to unify all the resident health records, digitalize healthy information and monitor chronic illness. To support different levels medical information development, DM has proposed electronic monitor solution for large disease prevention and cure.

DM-based medical solutions have many features including efficient, reliable, security, expandable, manageable, high cost-function ratio and good service.

DM-based medical solutions have achieved many applications and taken part in the national healthy information test projects, such as

(1) Beijing Tiantan Hospital: research on digital hospital development application on national software

(2) Beijing Xiehe Hospital: application of electronic medical record on national platform

(3) Shanghai Disease Prevention and Control Center: Ministry of Health electronic monitor system

(4) Wuxi Chinese Medicine Hospital: research on electronic medical record application on national software

(5) The First Affiliated Hospital of Zhejiang University: digital hospital on national software.

(6) Shanghai Zhabei Area: digital community healthy service system.

# Learning-Function-Augmented Inferences of Causalities Implied in Health Data⋆

JiaDong Zhang, Kun Yue, and WeiYi Liu

Department of Computer Science and Engineering,
School of Information Science and Engineering, Yunnan University,
Kunming, 650091, P.R. China
`jiadongzhang@yahoo.cn`

**Abstract.** In real applications of health data management, it is necessary to make Bayesian network (BN) inferences when evidence is not contained in existing conditional probability tables (CPTs). In this paper, we are to augment the learning function to BN inferences from existing CPTs. Based on the support vector machine (SVM) and sigmoid, we first transform existing CPTs into samples. Then we use transformed samples to train the SVM for finding a maximum likelihood hypothesis, and to fit a sigmoid for mapping outputs of the SVM into probabilities. Further, we give the approximate inference method of BNs with maximum likelihood hypotheses. An applied example and preliminary experiments show the feasibility of our proposed methods.

**Keywords:** Bayesian network inference, Support vector machine, Sigmoid, Learning function, Maximum likelihood hypothesis.

## 1 Introduction

In real applications of health data management, it is necessary to make probabilistic inference for diagnosis of disease according to patients' symptoms. Fortunately, the Bayesian network is a powerful uncertain knowledge representation and reasoning tool, which is widely studied and applied [1].

The basic task for probabilistic inference in a BN is to compute the posterior probability distribution for a set of query variables by search result, given some observed evidence. For example, in Fig. 1, we can deduce the probability distribution for *cholesterol standards* of somebody whose *age* is 60. However, in real applications, some queries are often submitted on arbitrary evidence values. Especially, we will have to consider the case when evidence values are not contained in existing conditional probability tables. For example, if we know John

⋆ This work was supported by the National Natural Science Foundation of China (No. 60763007), the Chun-Hui Project of the Educational Department of China (No. Z2005-2-65003), the Natural Science Foundation of Yunnan Province (No. 2005F0009Q), and the Cultivation Project for Backbone Teachers of Yunnan University.

Y. Ishikawa et al. (Eds.): APWeb 2008 Workshops, LNCS 4977, pp. 87–98, 2008.

| A | P(A) |
|---|---|
| $a_1=60$ | 0.3 |
| $a_2=70$ | 0.2 |
| $a_3=80$ | 0.3 |
| $a_4=90$ | 0.2 |

$A \rightarrow C$

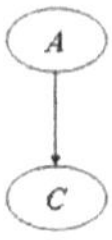

| $P(C|A)$ | $a_1=60$ | $a_2=70$ | $a_3=80$ | $a_4=90$ |
|---|---|---|---|---|
| $c_1=150$ | 2/3 | 0 | 0 | 0 |
| $c_2=170$ | 1/3 | 1 | 1/3 | 0 |
| $c_3=190$ | 0 | 0 | 2/3 | 1 |

| A | P(A) |
|---|---|
| $a_1=60$ | 0.3 |
| $a_2=70$ | 0.2 |
| $a_3=80$ | 0.3 |
| $a_4=90$ | 0.2 |

$A \rightarrow C$

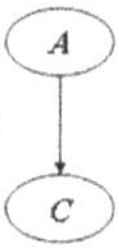

| $P(C|A)$ | $a_1=60$ | ... | $a_4=90$ | $a'=65$ |
|---|---|---|---|---|
| $c_1=150$ | 2/3 | ... | 0 | $h_{c1}(a')$ |
| $c_2=170$ | 1/3 | ... | 0 | $h_{c2}(a')$ |
| $c_3=190$ | 0 | ... | 1 | $h_{c3}(a')$ |

**Fig. 1.** A simple BN about *Age* (*A*) and *Cholesterol* (*C*)

**Fig. 2.** The BN with the maximum likelihood hypothesis for $a'$

is 65 years old, how to deduce the probability distribution for his *cholesterol standards*?

The ordinary inference with the BN (in Fig. 1) by search cannot answer this question, since there is no data about patients of 65 years old in the existing CPTs. For this problem, we discussed the inference method via learning maximum likelihood hypotheses from original samples when evidence values are not contained in existing CPTs [2]. A *maximum likelihood hypothesis* is an hypothesis that maximizes the probability of appearance of original samples [3]. Based on the maximum likelihood hypothesis, the posterior probability distribution for a set of query variables can be easily computed, given some observed evidence which is not contained in existing CPTs. For example, after finding maximum likelihood hypothesis, denoted as $h_C$ for *cholesterol standards*, the BN with $h_C$ corresponding to Fig. 1 is shown in Fig. 2. Based on $h_C$, the posterior probability distribution $h_C(a')$ for query variable $C$ can be computed, given evidence of $a' = 65$.

Actually, the probabilistic inference in a BN generally applies the conditional probability parameters instead of concerning the original samples. Thus, in this paper, we are to propose a method for learning the maximum likelihood hypothesis only from existing CPTs, i.e., augmenting the learning function to Bayesian network inferences.

Support vector machine is a new machine learning method based on the statistical learning theory. The support vector machine not only has solved certain problems in many learning methods, such as small sample, over fitting, high dimension and local minimum, but also has a fairly high generalization (forecasting) ability [4], [5]. In this paper, we are to extend general BNs by augmenting maximum likelihood hypotheses to make the inference done when evidence values are not contained in existing CPTs. Based on the SVM, we first transform existing CPTs into samples. Then we use the transformed samples to train the SVM for finding the maximum likelihood hypothesis with non-probabilistic outputs. It is required that outputs of the maximum likelihood hypothesis for BNs must be probabilities. Fortunately, sigmoid function is monotonic and takes on a threshold-like output, which can be adapted to give the best probability outputs [6]. Following, we can map the non-probabilistic outputs of the SVM into the probabilities via fitting a sigmoid using the transformed samples.

This approach is not only extending the expressive capability of a Bayesian network, but also finding a new application for SVMs.

Generally, the main contributions of this paper can be summarized as follows:

• Based on SVM and sigmoid function, we mainly give the method for obtaining maximum likelihood hypothesis from existing CPTs with respect to the evidence that are not contained in CPTs of BNs.

• We further give a Gibbs sampling algorithm for approximate probabilistic inference of BNs with maximum likelihood hypothesis.

• Aiming at real-world situations, an applied example is given and preliminary experiments are conducted to test the accuracy and feasibility of our proposed method are verified.

## 2    Related Work

As a graphical representation of probabilistic causal relationships, Bayesian networks (BNs) are effective and widely used frameworks [1], [7], [8]. A Bayesian network can be constructed by means of statistical learning from sample data [9], [10] and BNs have been used in many different aspects of intelligent applications [11]. Exact inference in large and connected networks is difficult [1], [8], and approximate inference methods are considered frequently, such as Monte Carlo algorithm [8]. Based on conditional independence, Pearl defined the Markov blanket, which describes the direct causes and direct effects given a certain node in a Bayesian network [1]. The discovery of Markov blanket is applied in the Monte Carlo algorithm for BN inferences [8].

As a novel statistic learning method, support vector machines have been paid wide attention recently [4], [5]. SVMs are based on the structural risk minimization principle from statistical learning theory [12]. The idea of structural risk minimization is to find a hypothesis which can guarantee the lowest error. SVM classification is to construct an optimal hyperplane, with the maximal marginal of separation between 2 classes [4], [5]. By introducing the kernel function, SVMs can handle non-linear feature spaces, and carry out the training considering combinations of more than one feature [4]. Furthermore, by training the parameters of an additional sigmoid function, the SVM outputs can be mapped into probabilities [6].

## 3    Learning the Maximum Likelihood Hypothesis from CPTs

As mentioned in Section 1, we cannot make inference with a general BN when the given evidence is not contained in existing CPTs. Moreover the probabilistic inference in a BN generally would not concern original samples. To conquer these difficulties, it is necessary to propose a method for learning the maximum likelihood hypothesis only from existing CPTs. Thus, the question is how we can implement such learning method from existing CPTs. In this section, we first introduce the SVM and sigmoid to learn the maximum likelihood hypothesis as the desired parameters in Bayesian networks, and then discuss the corresponding samples transformation from existing CPTs for obtaining parameters of the SVM and sigmoid.

## 3.1 Learning Parameters of Maximum Likelihood Hypothesis Based on the SVM and Sigmoid

It is known that SVMs are learning systems that use a hypothesis space of linear functions in a high dimensional feature space, trained with a learning algorithm from optimization theory that implements a learning bias derived from statistical learning theory [4], [5]. SVMs are quite suitable for the learning on small sample.

The unthresholded output of the standard linear 2-*classes* (i.e., *class* $\in$ $\{-1, 1\}$) SVM [12] is

$$f(x) = (w \bullet x) + b, \tag{1}$$

where

$$w = \sum_{i=1}^{n} y_i \alpha_i x_i \tag{2}$$

and

$$b = y_i - \sum_{i=1}^{n} y_i \alpha_i (x_i \bullet x_j), \forall j \in \{0 < \alpha_j < c\}. \tag{3}$$

Obviously, the output $f$ is not a probability, where $f = f(x)$. To map the SVM outputs into probabilities, we adopt the method that applies a parametric model to fit the posterior $P(y = 1|f)$ (i.e., $P(class|input)$) directly. And the parameters of the model should be adapted to give the best probability outputs. The probability $P(y = 1|f)$ should be monotonic in $f$, since the SVM is trained to separate most or all of positive examples from negative examples. Moreover $P(y = 1|f)$ should lie between 0 and 1. Bayes' rule suggests using a parametric form of a sigmoid [6]:

$$P(y = 1|f) = \frac{1}{1 + \exp(\beta f + \gamma)}. \tag{4}$$

This means that the maximum likelihood hypothesis $h = P(y = 1|f)$ can be computed by the following method. First the non-probabilistic output $f$ of the SVM can be computed based on equation (1). Then the non-probabilistic output of the SVM can be mapped into the probability $P(y = 1|f)$ based on equation (4). The computation details are interpreted by the following example.

**Example 1.** Let us consider the Bayesian network with conditional probability tables shown in Fig. 3, where variables $A$, $C$, $F$ and $H$ denote *age, cholesterol, family history* of *hypertension disease, hypertension disease* respectively, and $H = \{0, 1, 2, 3\}$ indicates that the blood pressure of a patient is normal, 1-*phase*, 2-*phase* and 3-*phase* respectively.

If the inference is on $C$ under the evidence of $A = a' = 65$, it is necessary to concern $P(C|A = a')$, which is not included in Fig. 3 and can be obtained by

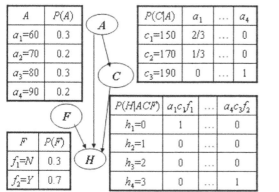

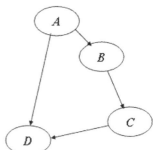

**Fig. 3.** A Bayesian network

**Fig. 4.** The structure of the Bayesian network $G$

our proposed method. We are to obtain the posterior probability distribution $h_C(a')$ (i.e., $P(C|A = a')$) by taking $C$ as the target.

Considering $C = c_1 = 150$, first we set the target to 1 for all samples with $C = c_1 = 150$, and $-1$ for others. Then suppose that we can obtain $w = 0.2, b = -12.4$ by training the SVM, $f(a') = (w \bullet a') + b = 0.6$ can be computed based on equation (1). Further suppose that we can obtain $\beta = -2.3, \gamma = 0$ by fitting the sigmoid, the non-probabilistic output $f(a') = 0.6$ can be mapped into the probability $P(y = 1|f(a')) = 0.8$ based on equation (4).

Therefore, we have $h_{c_1}(a') = P(y = 1|f(a')) = 0.8$ based on equations (1) and (4). Analogously, we have $h_{c_2}(a') = 0.2$ and $h_{c_3}(a') = 1$. After normalization, we can obtain $h_{c_1}(a') = 0.4$, $h_{c_2}(a') = 0.1$, and $h_{c_3}(a') = 0.5$.

Similarly, by computing maximum likelihood hypothesis, we can obtain the parameters for each node in Fig. 3. Suppose $a' = 65, c' = c_1 = 150$ and $f' = N$, the parameters of maximum likelihood hypothesis $h_H$ on $H$ can be obtained as follows:

$$h_{h_1}(a'c'f') = 0.2, h_{h_2}(a'c'f') = 0.6, h_{h_3}(a'c'f') = 0.1, \text{ and } h_{h_4}(a'c'f') = 0.1.$$

Till now, we have expounded the method for learning the maximum likelihood hypothesis for the linear case. For the nonlinear case, we only need make the following transformation. First the data is mapped into some other dot product space $F$, called the feature space, by introducing a nonlinear map $\phi : R^m \rightarrow F$. Then $x$ and dot product $(x \bullet x')$ are replaced by $\phi(x)$ and dot product $(\phi(x) \bullet \phi(x'))$ respectively, in which $x$ and $x'$ are 2 given samples. In order to compute dot products of the form $(\phi(x) \bullet \phi(x'))$, we employ kernel representations of the form $K(x, x') = (\phi(x) \bullet \phi(x'))$ [4].

Thus whatever cases mentioned above, we can obtain the maximum likelihood parameter $h(x_i) = P(y = 1|f(x_i))$ for each given $x_i$ that may be an evidence value and is not contained in existing CPTs. The maximum likelihood parameter $h(x_i) = P(y = 1|f(x_i))$ is regarded as the corresponding entry (i.e., $P(y = 1|x_i)$)

in conditional probability tables. Therefore, inferences taking $x_i$ as the evidence can be made.

In the above example, it is known SVM and sigmoid need be trained and be fitted respectively for obtaining parameters $w, b, \beta$ and $\gamma$. And the probabilistic inference in a Bayesian network generally would not concern the original samples. Thus we are to discuss samples transformation methods from CPTs for training the SVM and fitting the sigmoid, which will be introduced in subsection 3.2 and subsection 3.3 respectively.

## 3.2    Transforming CPTs into Samples for Training the SVM

In this subsection, we are to transform existing conditional probability tables into training samples to obtain the parameters $w$ and $b$.

In a Bayesian network, each conditional probability table corresponds to a variable, denoted as $y$, and the set of parents of $y$, denoted as $x$.

For each combinational state $x_j$ of $x$, it corresponds to a conditional probability distribution $P(y|x_j)$,

$$P(y = y_k|x_j) = p_{jk}, k = 1, 2, \cdots, \sum_k p_{jk} = 1,$$

where $y_k$ is a state of $y$.

Based on the conditional probability distribution $P(y|x_j)$, we can obtain a simple for the combinational state $x_j$ using the inverse transform algorithm (see Algorithm 1) [13]. The basic idea is immediately given as follows. First, a random number is generated. Then, the value of $y$ is determined based on the generated random number and the conditional probability distribution $P(y|x_j)$.

**Algorithm 1.** Inverse transform algorithm
**Input:**
    $P(y|x_j)$: a conditional probability distribution corresponding to the combinational state $x_j$ of $x$.
**Output:**
    Samples generated based on $P(y|x_j)$.
    **Step 1:** Generate a random number $r$, where $r$ is uniformly distributed over $(0, 1)$.
    **Step 2:** Determine the values of $y$:

$$y = \begin{cases} y_1 & r \leq p_{j1} \\ y_2 & p_{j1} < r \leq p_{j1} + p_{j2} \\ \vdots & \vdots \end{cases}$$

By repeated executions of the Algorithm 1, sampling can be done for the combinational state $x_j$ of $x$. Similarly, sampling can be done for other combinational states of $x$ based on Algorithm 1, and all samples are denoted as set $\{(x_i, y_i)\}$. Following, an example will be shown to illustrate the application of Algorithm 1.

**Table 1.** The training set $\{(x_i, y_i)\}$ corresponding to the CPT of $C$

| Age $(A)$ | Cholesterol $(C)$ |
|-----------|-------------------|
| 60        | 170               |
| 60        | 150               |
| $\vdots$  | $\vdots$          |
| 90        | 190               |
| 90        | 190               |

**Table 2.** The training set $\{(f_i, t_i)\}$ for the class $H = h_1 = 0$ of $H$

| $f$ | $t$ |
|-----|-----|
| $f(60, 150, N)$ | 1 |
| $f(60, 150, Y)$ | 3/4 |
| $\vdots$ | $\vdots$ |
| $f(90, 190, N)$ | 0 |
| $f(90, 190, Y)$ | 0 |

**Example 2.** On the BN in Fig. 3, considering the conditional probability table of variable $C$, $\{A\}$ is the set of parents of $C$. Sampling can be done based on the conditional probability table. For state $a_1 = 60$ of $A$, the corresponding conditional probability distribution $P(C|a_1)$ is as follows:

$$P(C = c_1|a_1) = 2/3, P(C = c_2|a_1) = 1/3, P(C = c_3|a_1) = 0.$$

Following steps are executed repeatedly:

1. Generate a random number $r$ and suppose $r = 0.8$.
2. The result will be $C = c_2$, since $2/3 < 0.8 < 2/3 + 1/3$.

Similarly, sampling can be done for states $A = a_2$, $A = a_3$ and $A = a_4$. The training set $\{(x_i, y_i)\}$ is shown in Table 1.

Now we have obtained the training set $\{(x_i, y_i)\}$ for training the SVM from the existing conditional probability tables, so the parameters $w$ and $b$ can be solved by the conventional methods [12]:

$$
\begin{cases}
\min_{\alpha} \frac{1}{2} \sum_{i=1}^{n} \sum_{j=1}^{n} y_i y_j \alpha_i \alpha_j (x_i \bullet x_j) - \sum_{i=1}^{n} \alpha_i \\
s.t. \sum_{i=1}^{n} y_i \alpha_i = 0 \\
\quad 0 \le \alpha_i \le c, \ i = 1, \cdots, n
\end{cases}
\tag{5}
$$

For example, let $x = \{(1,3), (4,3), (2,4), (1,1), (4,1), (2,0)\}$ be a set of samples, and $y = \{1, 1, 1, -1, -1, -1\}$ be a set of corresponding targets. By solving the optimization problem (5), we can obtain $\alpha = (0.5, 0, 0, 0.5, 0, 0)$. Based on equations (2) and (3), we obtain $w = (0, 1)$ and $b = -2$ respectively.

### 3.3 Transforming CPTs into Samples for Fitting the Sigmoid

The parameters $\beta$ and $\gamma$ can be fitted based on maximum likelihood estimation from a training set $\{(f_i, t_i)\}$. Similarly, we can transform conditional probability tables into the training set $\{(f_i, t_i)\}$.

When considering some class (i.e., a state of $y$), we set the target to 1 for the class, and $-1$ for others. Thus, each combinational state $x_i$ of $x$ corresponds to an entry $P(y = 1|x_i)$ in the conditional probability table. $P(y = 1|x_i)$ can be transformed into an element $(f_i, t_i)$ of the training set by the following method.

- In subsection 3.2, we obtain the parameters $w$ and $b$, so $f_i$ can be obtained based on the following equation:

$$f_i = f(x_i). \tag{6}$$

- $t_i$ is the target probability defined simply as follows:

$$t_i = P(y = 1|x_i). \tag{7}$$

This means that conditional probability tables can be transformed into the training set $\{(f_i, t_i)\}$ by the following method. First the non-probabilistic output $f(x_i)$ of the SVM is taken as $f_i$ by equation (6). Then corresponding entry $P(y = 1|x_i))$ in conditional probability tables is taken as $t_i$ by equation (7). The details of transforming are interpreted by the following example.

**Example 3.** On the BN in Fig. 3, considering conditional probability table of variable $H$, given the class $H = h_1 = 0$, each combinational state $(a', c', f')$ of $\{A, C, F\}$ corresponds to an entry $P(h_1|a'c'f')$. For combinational state $(a_1 = 60, c_1 = 150, f_1 = N)$, the entry $P(h_1|a_1c_1f_1) = 1$ can be transformed into the sample $(f(a_1, c_1, f_1), 1)$, where $f(a_1, c_1, f_1)$ is easy to be computed based on the method presented in subsection 3.2. Analogously, we can map other combinational states of $\{A, C, F\}$ into samples. Therefore, the training set for the class label $H = h_1 = 0$ can be obtained, which is shown in Table 2. Similarly, we can obtain the training set for any class label of any variable.

Consequently, the sigmoid can be fitted by the following method.

The parameters $\beta$ and $\gamma$ are found by minimizing the negative log likelihood of the training set $\{(f_i, t_i)\}$, which is a cross-entropy error function [6]:

$$\min_{\beta, \gamma} - \sum_{i=1}^{n} t_i \log(p_i) + (1 - t_i) \log(1 - p_i), \tag{8}$$

where

$$p_i = \frac{1}{1 + \exp(\beta f_i + \gamma)}.$$

For example, let $\{(2, 1), (1, 0.9), (0, 0.5), (-1, 0.05), (-2, 0)\}$ be a training set. By solving the optimization problem (8), we can obtain $\beta = -2.7$ and $\gamma = 0.1$.

## 4   Approximate Inference of Bayesian Networks with Maximum Likelihood Hypothesis

The intractability of exact inference of interval probability is obvious. Based on the property of Markov blankets, we consider adopting Gibbs sampling algorithm [7] for approximate inference with the Bayesian network. A Markov blanket $MB(x)$ [1] of a node $x$ in a Bayesian network is any subset $S(x \notin S)$ of nodes for which $x$ is independent of $U - S - x$ given $S$, where $U$ is a finite set of nodes.

We extend the Gibbs-sampling probabilistic algorithm to the Bayesian network with maximum likelihood hypotheses in Algorithm 2 [2].

**Algorithm 2.** Approximate inference in Bayesian networks with the learning function
**Input:**
  $BN$: a Bayesian network with conditional probability parameters and maximum likelihood hypothesis
  $\vec{Z}$: the nonevidence nodes in $BN$
  $\vec{E}$: the evidence nodes in $BN$
  $x$: the query variable
  $\vec{e}$: the set of values of the evidence nodes $\vec{E}$
  $n$: the total number of samples to be generated
**Output:** The estimates of $P(x|e)$.
Variables:

  $\vec{z}$: the set of values of the nonevidence nodes $\vec{Z}$
  $N_x(x_i)(i = 1, \cdots, n)$: a vector of counts over probabilities of $x_i$, where $x_i(i = 1, \cdots, n)$ is the value of $x$.
  **Step 1.** Initialization:
  $\vec{z} \leftarrow$ random values;
  $\vec{e} \leftarrow$ evidence values of $\vec{E}$;
  $N_x(x_i) \leftarrow 0 \ (i = 1, \cdots, n)$
  **Step 2.** For $i \leftarrow 1$ to $n$ do
  (1) Compute the probability values $P(x_i|MB(x))$ of $x$ in the next state where $MB(x)$ is the set of the current values in the Markov blanket of $x$.
  • If evidence values are in the existing CPTs, then $P(x_i|MB(x))$ can be obtained by searching in the conditional probability tables.
  • If evidence values are not in the existing CPTs, then $P(x_i|MB(x))$ can be obtained by computing the maximum likelihood hypothesis.
  (2) Generate a random number $r$, where $r$ is uniformly distributed over $(0, 1)$, we determine the values of $x$:

$$
x = \begin{cases} x_1 & r \le P(x_1|MB(x)) \\ x_2 & P(x_1|MB(x)) < r \le P(x_1|MB(x)) + P(x_2|MB(x)) \\ \vdots & \vdots \end{cases}
$$

  (3) Count:
  If $x = x_i$ then $N_x(x_i) \leftarrow N_x(x_i) + 1$
  **Step 3.** Estimate $P(x|e)$:
  $P(x_i|e) \leftarrow N_x(x_i)/n$.

# 5   An Applied Example and Experimental Results

## 5.1   An Applied Example

The diagnosis of disease according to patients' symptoms using Bayesian networks have been studied. Let us consider the database concerning *America*

**Table 3.** The training set $\{(x_i, y_i)\}$ corresponding to the CPT of $B$

| A | B |
|---|---|
| 40 | 100 |
| 40 | 120 |
| ⋮ | ⋮ |
| 70 | 160 |

**Table 4.** The training set $\{(f_i, t_i)\}$ for the class $B = b_1 = 100$ of $B$

| f | t |
|---|---|
| -0.3 | 0.3 |
| -0.5 | 0.1 |
| -0.7 | 0.1 |
| -0.9 | 0.1 |

*Cleveland heart disease diagnosis* downloaded from UCI machine learning repository [14]. In order to focus on the idea of our proposed method, we only consider the 4-attributes of *age*, *resting blood pressure* (mm Hg), *serum cholesterol* (mg/dl) and *diagnosis* of *heart disease* (the predicted attribute), which are denoted as variables $A$, $B$, $C$ and $D$ respectively.

By means of the Scoring&search based algorithm for constructing Bayesian networks [1], we can obtain the structure of the Bayesian network $G$ and CPTs (shown in Fig. 4), in which the CPTs are leaved out for space limitation. The maximum likelihood hypotheses of $h_B$ and $h_D$ can be obtained by our proposed method.

For the maximum likelihood hypothesis $h_B$, considering the conditional probability table corresponding to variable $B$, the training set $\{(x_i, y_i)\}$ can be obtained at first based on Algorithm 1. The obtained training set is shown in Table 3 including 50 entries.

Considering $B = b_1 = 100$, we first set the target to 1 for all samples (shown in Table 3) with $B = b_1 = 100$, and $-1$ for others. Then by solving the optimization problem (5), the parameters $w = -0.02$ and $b = 0.5$ can be obtained based on equations (2) and (3) respectively. Thus, the non-probabilistic output $f(x)$ of the SVM can be computed based on the following equation:

$$f(x) = -0.02x + 0.5, \tag{9}$$

where $x$ is any given evidence value of *age*. Further, the training set $\{(f_i, t_i)\}$ can be obtained based on equations (6), (7) and (9), shown in Table 4. By solving the optimization problem (8), we can obtain the parameters $\beta = -2.5$ and $\gamma = 0.4$. Therefore, based on equations (4) and (9), we can obtain the maximum likelihood hypothesis $h_{b_1}$,

$$h_{b_1} = P(y = 1|f) = \frac{1}{1 + \exp(-2.5f + 0.4)} = \frac{1}{1 + \exp(0.05x - 0.85)}.$$

Analogously, we can obtain $h_{b_2}$, $h_{b_3}$ and $h_{b_4}$.

Similarly, we can obtain the maximum likelihood hypothesis $h_D$ for variable $D$.

Based on the Bayesian networks with these maximum likelihood hypotheses, we can apply Algorithm 2 to make the inference done when evidence values are not contained in existing conditional probability tables. For example, considering the query $P(D|A = 65, C = 350)$ applied to the network, $P(D|A = 65, C = 350)$

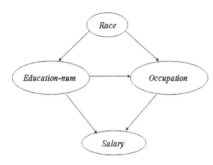

**Fig. 5.** The structure of the BN for 4-attributes of *Adult*

**Table 5.** Errors of conditional probability distribution for removing combinational states of variable *occupation*

| Race | Education-num | Error |
|---|---|---|
| Amer-Indian-Eskimo | 5 | 0.1213 |
| Amer-Indian-Eskimo | 15 | 0.1183 |
| Asian-Pac-Islander | 9 | 0.0593 |
| Black | 3 | 0.1220 |
| Black | 13 | 0.1443 |
| Other | 6 | 0.1702 |
| White | 1 | 0.1922 |
| White | 9 | 0.0986 |

can be computed, since the posterior probability distributions $h_B(65)$ and $h_D(65, 350)$ can be easily computed based on $h_B$ and $h_D$ respectively.

## 5.2   Experimental Results

To verify the feasibility of our proposed methods in this paper, we tested their accuracy and effectiveness for learning maximum likelihood hypothesis from the existing CPTs. Our experiments involved the samples *Adult* downloaded from the UCI machine learning repository [14], and we only consider the attributes *race*, *education-num*, *occupation* and *salary* (the predicted attribute) of *Adult*. We first constructed the Bayesian network (see Fig. 5) and corresponding CPTs for the 4-attributes using the Bayesian network learning tool - PowerConstructor [9]. Then for some variable, we removed some combinational states of parents of the variable, and computed the corresponding conditional probability distribution for these combinational states based on our proposed method. Consequently, for each removed combinational state, errors were obtained by the absolute values of the maximum likelihood hypothesis learned by our proposed method minus the corresponding conditional probabilities in the BN learned from the original sample data.

We specifically considered 8 different combinational states of *race* and *education-num* for *occupation*. Errors are given in Table 5. Then, the general maximum, minimal and average errors are 0.1922, 0.0593 and 0.1283 respectively.

From above results of the accuracy tests, we can conclude that our proposed method is effective and accurate to a great extent, and thus can be applied into corresponding inferences in disease diagnosis.

## 6   Conclusion and Future Work

In this paper, we augment learning function to BN inferences from existing CPTs. Based on the SVM and sigmoid, we first transform existing conditional probability tables into samples, then train the SVM for finding the maximum likelihood hypothesis, and map the outputs of the SVM into the probabilities via fitting a sigmoid. Further, we give the approximate inference method of Bayesian networks with maximum likelihood hypothesis. Finally, an applied example is given and experimental results show that our proposed methods are feasible.

Our methods also raise some other interesting research issues. For example, the integration and fusion of multiple BNs with maximum likelihood parameters can be further studied based on the ideas presented in this paper.

# References

1. Pearl, J.: Probabilistic reasoning in intelligent systems: network of plausible inference. Morgen Kaufmann publishers, Inc., San Mates, California (1998)
2. Liu, W.Y., Yue, K., Zhang, J.D.: Augmenting Learning Function to Bayesian Network Inferences with Maximum Likelihood Parameters. Technical Report, Yunnan University (2007)
3. Mitchell, T.M.: Machine Learning. McGraw-Hill Companies, Inc., New York (1997)
4. Burges, C.J.C.: A tutorial on support vector machines for pattern recognition. Data Mining and Knowledge Discovery 2(2), 121–167 (1998)
5. Chang, C., Liu, C.J.: Training v-support vector classifiers: theory and algorithms. Neural Computing 13(9), 2119–2147 (2001)
6. Platt, J.C.: Probabilistic Outputs for Support Vector Machines and Comparisons to Regularized Likelihood Methods. In: Advances in Large Margin Classifiers. MIT Press, Cambridge (1999)
7. Heckerman, D., Wellman, M.P.: Bayesian networks. Communication of the ACM 38(3), 27–30 (1995)
8. Russel, S.J., Norving, P.: Artificial intelligence - a modern approach. Pearson Education Inc., Publishing as Prentice-Hall (2002)
9. Cheng, J., Greiner, R., Kelly, J., Bell, D., Liu, W.: Learning Bayesian network from data: an information-theory based approach. Artificial Intelligence 137(2), 43–90 (2002)
10. Buntine, W.L.: A guide to the literature on learning probabilistic networks from data. IEEE Trans. on Knowl. Data Eng. 8(2), 195–210 (1996)
11. Heckerman, D., Mandani, A., Wellman, M.P.: Real-world applications of Bayesian networks. Communications of the ACM 38(3), 25–30 (1995)
12. Vapnik, V.: Statistical learning theory. John Wiley and Sons, Inc., New York (1998)
13. Ross, S.M.: Simulation, 3rd edn. Academic Press, Inc., London (2002)
14. Asuncion, A., Newman, D.J.: UCI Machine Learning Repository. Irvine, CA: University of California, Department of Information and Computer Science (2007), http://www.ics.uci.edu/~mlearn/MLRepository.html

# Support Vector Machine for Outlier Detection in Breast Cancer Survivability Prediction

Jaree Thongkam, Guandong Xu, Yanchun Zhang, and Fuchun Huang

School of Computer Science and Mathematics
Victoria University, Melbourne VIC 8001, Australia
{jaree,xu,yzhang,fuchun}@csm.vu.edu.au

**Abstract.** Finding and removing misclassified instances are important steps in data mining and machine learning that affect the performance of the data mining algorithm in general. In this paper, we propose a C-Support Vector Classification Filter (C-SVCF) to identify and remove the misclassified instances (outliers) in breast cancer survivability samples collected from Srinagarind hospital in Thailand, to improve the accuracy of the prediction models. Only instances that are correctly classified by the filter are passed to the learning algorithm. Performance of the proposed technique is measured with accuracy and area under the receiver operating characteristic curve (AUC), as well as compared with several popular ensemble filter approaches including AdaBoost, Bagging and ensemble of SVM with AdaBoost and Bagging filters. Our empirical results indicate that C-SVCF is an effective method for identifying misclassified outliers. This approach significantly benefits ongoing research of developing accurate and robust prediction models for breast cancer survivability.

**Keywords:** Outlier Detection System, C-Support Vector Classification Filter (C-SVCF), Breast Cancer Survivability.

## 1   Introduction

Breast cancer survivability prediction tasks are commonly categorized as a binary classification of 'dead' or 'alive' at a point in time. Breast cancer survivability data in hospital databases is commonly collected without any specific research purpose [1] [2]. However, this kind of data may contain records that do not follow the common rules, and affect the model's performance. For example, patients who have breast cancer in stage I and are aged less than 30 years old, should be categorized as 'alive'. However, these patients have been categorized as 'dead' in the data set, due to having died of other illnesses. In these cases we assume data as outliers.

Currently, outliers in data commonly result in overfitting problems in the learning model to induct an overly specific hypothesis to fit the training data well, but performs poorly on unseen data [3]. Besides, most inductive learning aims to form a generalization from a set of training instances so that classification accuracy on previously unseen instances is maximized. As a result, the maximum achievable accuracy depends on the quality of the data and the appropriateness of the biases of the chosen

Y. Ishikawa et al. (Eds.): APWeb 2008 Workshops, LNCS 4977, pp. 99–109, 2008.
© Springer-Verlag Berlin Heidelberg 2008

learning algorithm for the data [4]. Therefore, finding outliers is the most important step in the practice of data mining tasks.

Three common outlier handling approaches in data include robust algorithms, outlier filtering, and correcting outlier instances [5]. First, robust algorithms are used to build a complex control mechanism to avoid overfitting training data and generalize well in unseen data [6]. Second, outlier filtering techniques employ the learning algorithm to identify and eliminate potential outliers from mislabeled instances in the training set [4]. Finally, outlier correction methods are built upon the assumption that each attribute or feature in the data is correlated with others, and can be reliably predicted [7]. However, outlier correction methods are usually more computationally expensive than robust algorithms and outlier filtering techniques, and it is unstable in correcting and cleaning unwanted instances [4]. Therefore in this paper, outlier filtering techniques are chosen to identify and remove misclassified instances.

The outlier filtering technique was motivated by the need to find techniques for improving the performance of classifiers [4] [8] [9]. Since classification techniques have poor results when data contain the specific value of the outliers. The filtering methods are commonly used to filter the outliers in the training set and improve the quality of the training data set. Several research studies have employed outlier filtering techniques for identifying and eliminating subjects with misclassified instances in training sets. For example, Verbaeten and Assche [3] utilized inductive logic programming (ILP) and a first order decision tree algorithm to construct the ensembles. Their techniques started with the outlier-free data set followed by adding the different levels of classification outlier in the outlier-free data set, and evaluated the effectiveness using a decision tree. Their results showed that the accuracy of decision tree was decreased rapidly after increasing the levels of noise. Furthermore, Blanco, Ricket and M-Merino [10] combined multiple dissimilarities and support vector machine (SVM) to filter the spam massages in processing e-mail. Their results showed that the combination of multiple dissimilarities and SVM was better than dissimilarity alone.

*Support Vector Machine* is emerging as a popular technique in machine learning [11]. This approach is a novel classification technique, and based on neural network technology using statistical learning theory [12]. Moreover, there are a number of different variants of support vector machines. The simplest one of them is the C-Support Vector Classification (C-SVC) algorithm, which suits binary class classification problems [11]. For instance, Yin, Yin, Sun and Wu [13] utilized C-SVC with the radial basis function to identify classification problems in handwritten Chinese characters. Their results showed that C-SVC with radial basis function has the highest accuracy for their predicting tasks.

Although several research studies have used support vector machine methods for filtering spam email and pattern recognition, few research studies employ C-SVC to filter the outlier in the training data. In this paper, we propose a C-SVC as the filter (namely C-SVCF) for identifying and eliminating misclassified instances in order to improve classification accuracies produced by learning algorithms in the breast cancer survivability data set. In experiments, 5% up to 20% of misclassified trained instances were randomly selected to remove from the training data, and classification performance in terms of accuracy and the AUC score was evaluated using 10-fold cross-validation via various learning algorithms.

This paper is organized as follows. Section 2 reviews the outlier filtering approach and the C-SVCF algorithm. Section 3 presents the methodologies and research experiments used in this paper. Next, experiment results and discussions are presented in section 4. The conclusion and suggestions for future work are given in Section 5.

## 2   Basic Concept of Outlier Filtering

In this section, the outlier filtering framework is introduced, and a C-SVCF algorithm is proposed to identify and eliminate the misclassified instances in the breast cancer survivability data set.

### 2.1   Outliers Filtering Framework

In the medical databases, raw data may contain instances that do not comply with the model behavior. These instances are referred to as outliers [11]. Outliers may be detected using statistical tests that assume a distribution or probability model for the data, or using distance measures where instances that are a substantial distance from any other cluster are considered outliers. The idea of eliminating instances to improve the performance of classifiers has been a focus in pattern recognition and instance-based learner [4] [11].

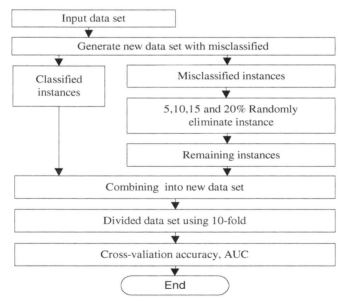

**Fig. 1.** Outlier filtering framework

In practice, outlier filtering frameworks are generally started with the original data set without outliers. Then, the different levels of outliers are added into the original and evaluated through learning algorithms to identify and remove misclassified instances such as [3], [14] and [15]. In contrast, our framework started from data with outliers, and reduced the outlier by applying the C-SVCF algorithm to identify and eliminate those outliers. For example, the algorithm marks an instance as mislabeled if it is classified wrongly, whereas the algorithm marks an instance as correct if it is classified correctly. Following this we randomly eliminated the misclassified instances by 5% of the original data set each time until 20%, to simulate the ability of C-SVCF. The outlier filtering framework is introduced in Fig.1.

The aims of this outlier filtering framework are to find the techniques which identify less outliers and high performance. Since we have reduced the classification task with misclassified instances as a deterministic task, the percentage of eliminated misclassified instances should improve the prediction performance in any classification algorithm. Experiment results are demonstrated in Section 4.1 to verify this point.

## 2.2  C-Support Vector Classification Filter (C-SVCF)

C-Support Vector Classification (C-SVC) [12] is a binary classification in the Support Vector Machine (SVM) [16] family, which is a new generation learning system based on recent advances in statistical, machine learning and pattern recognition [13]. In relation to medical mining, Yi and Fuyong [17] utilized C-SVC to identify breast cancer diagnosis attributes in an unbalanced training set, and searched for suitable attributes to reduce the computational costs. Their results showed a significant improvement of accuracy (up to 97.1%) when they applied the C-SCV algorithm with the polynomial function to select the appropriate attributes.

In this paper, we propose the C-SVCF algorithm which uses C-SVC with the radial basis kernel function as an instances filtering method. The C-SVCF algorithm is given in Fig. 2.

| C-SVCF Algorithm |
| --- |
| Input : $D$ (TrainingData with n instances) |
| Output: $F$ (FilteredData) |
|         $O$ (OutlierData) |
| $T \leftarrow$ C-SVC(TrainingData) |
| $F$: empty set |
| $O$: empty set |
| for i = 1 to n |
|     if instance $\in$ $T$ then |
|         insert to $F$ |
|    else |
|         insert to $O$ |
|    end if |
|   next |
| Output: $F,O$ |

**Fig. 2.** C-Support Vector Classification Filter (C-SVCF) Algorithm

Fig. 2 shows that the C-SVCF algorithm is used to identify and eliminate outliers from training data. After running the C-SVCF algorithm, the result is a set of filtered data and a set of outlier data. Although the kernel functions are an important parameter of SVM in some application fields such as text classification and number recognition [13], the problem of designing the kernel function is out of the scope of this paper. Besides, although we have done the experiments using different kernel functions, using the radial basis kernel function in C-SVC type leads to the best result.

# 3 Methodologies and Evaluation Methods

In this study, we investigate the capability of using C-SVCF as an indemnification and elimination method to improve performance of the breast cancer survivability model. In this section we first present the methodologies of breast cancer survivability used in the experiment. Following this, evaluation methods are presented in terms of accuracy and AUC score.

## 3.1 Breast Cancer Survivability Data Set

The breast cancer survivability data was obtained from Srinagarind Hospital in Thailand. The data include patient information, and the choice of treatments of patients who were diagnosed with breast cancer from 1990 to 2001. The breast cancer survivability data consists of 2,462 instances and 26 attributes. After studying descriptive statistics, the results showed that some attributes have more than 30% of missing values, while some attributes have only one value. The reason is that some patients have been diagnosed in Srinagarind, but received treatments in other hospitals. Therefore, the final data set consists of 736 instances, 11 attributes and a class or binary attribute. Accordingly, a binary attribute was selected in which patients surviving less than 60 months are coded as 0 (Dead) and more than 60 months as 1 (Alive). Therefore, the class attribute includes both 'Dead' and 'Alive'. The 'Dead' class is composed of 394 instances, and the 'Alive' class is composed of 342 patients. The attribute list is presented in Table 1.

**Table 1.** Input attributes of breast cancer data set

| No | Attributes | Types |
|----|-----------|-------|
| 1 | Age | Number |
| 2 | Marital Status | Category(3) |
| 3 | Occupation | Category(27) |
| 4 | Basis of diagnosis | Category(6) |
| 5 | Topography | Category(9) |
| 6 | Morphology | Category(14) |
| 7 | Extent | Category(4) |
| 8 | Stage | Category(4) |
| 9 | Received Surgery | Category(2) |
| 10 | Received Radiation | Category(2) |
| 11 | Received Chemo | Category(2) |
| 12 | Survivability (Class attribute) | Category(2) |

## 3.2 Evaluation Methods

Two evaluation methods, accuracy and an area under ROC curve, are used to evaluate the performance of the filtering method in this paper. These evaluation methods are commonly used in evaluating the performance of learning algorithms.

**Accuracy.** Accuracy is the basic method for presenting the performance of the classifier related to generalization error [11]. In this paper, accuracy is the percentage of the correct predictions when compared with actual classes among the test set. This accuracy is defined in Equation 1.

$$accuracy = \frac{TP + TN}{TP + FP + TN + FN} \tag{1}$$

where: $TP$ represents the true positive value which is the number of correct predictions in the 'Alive' class. On the other hand, $FP$ represents the false positive value which is the number of incorrect predictions in the 'Alive' class. $TN$ represents the true negative value which is the number of correct predictions in the 'Dead' class. In contrast, $FN$ represents false negative value which is the number of incorrect predictions in the 'Dead' class. In this paper, we use the estimation of the performance using a 10-fold cross-validation approach to reduce the bias associated with the random sampling strategy [18] [19].

**Area under the Receiver Operating Characteristic curve.** An Area under the ROC Curve (AUC) is traditionally used in medical diagnosis system. Recently, this has been proposed as an alternative measure for evaluating the predictive ability of learning algorithms, and measure of test accuracy. Besides, Huang and Ling [20] demonstrated that AUC is a better measurement method than accuracy. An AUC is equivalent to the probability that randomly selected instances of one class have a smaller estimated probability of belonging to the other classes than one randomly selected from the other classes [21]. Moreover, it provides an approach for evaluating models that is based on an average which graphically interprets the performance of the decision-making algorithm with regard to the decision parameter by plotting the true positive rate against the false positive rate [22], [23] (see Fig. 3).

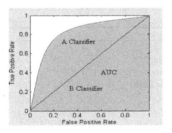

**Fig. 3.** The area under the ROC curve (AUC)

AUC usually has scores of between 0 and 1 for classifier performance. Furthermore, the interpretation of AUC results is easy to understand [20]. For example, Fig. 3 shows the areas under two ROC curves, $A$ and $B$ classifiers. The AUC score of $A$ classifier is larger than $B$ classifier, therefore, $A$ classifier has better performance than $B$ classifier. Several research studies have utilized AUC for comparing the classifiers' performance. While Jiang [24] employed the average AUC to analysis the optimal linear in artificial neural network (ANN) output, the current study utilizes the average AUC as a performance selection criterion of the filtering method in classification task.

## 4  Results and Discussions

In this section, the WEKA version 3.5.6 [25] and LIBSVM [26] data mining environment are selected to evaluate the proposed method. The WEKA environment is a well-defined framework, and offers a variety of classification algorithms for identifying outliers in a data set. Experiments are performed on a breast cancer survivability data set from Srinagarind hospital in Thailand. We first present the number of misclassified instances using C-SVCF, and then compare it with AdaBoost, Boosting and SVM ensembles (see Section 4.1). In Section 4.2 and 4.3 the effectiveness of outlier detection and elimination is evaluated by randomly selecting instances from the outlier data set by 5% each time, and comparing the classification performance with AdaBoost, Boosting and SVM ensembles filtering using seven learning algorithm including C4.5, conjunctive rule, Naïve Bayes, nearest-neighbour classifier (NN-classifier), random committee, random forests and radial basis function network (RBFNetwork), respectively.

### 4.1  Outliers from Misclassified Instances

In these experiments, the number of misclassified instances using C-SVCF is compared with other outlier filtering techniques including AdaBoost, Bagging and SVM ensemble filtering. We use the decision stump as a weak learner in the AdaBoost algorithm [25], and employ the fast decision tree learner (called REPTree) as weak learner in Bagging algorithm. Furthermore, we employ SVM ensemble by using C-SVC with radial kernel function as base learner in AdaBoost (called AdaBoostSVM). In addition, C-SVC with radial kernel function as base learner in Bagging (called BaggingSVM) is utilized to compare with the proposed method. The results of the misclassified instances are shown in Table 2.

**Table 2.** The number of misclassified instances

| Filters | Number of classified | | Number of misclassified | | Total number of misclassified | Percentage of misclassified |
|---|---|---|---|---|---|---|
| | 'Dead' | 'Alive' | 'Dead' | 'Alive' | | |
| AdaBoost | 316 | 185 | 78 | 157 | 235 | 31.93 |
| Bagging | 283 | 241 | 111 | 101 | 212 | 28.80 |
| AdaBoostSVM | 338 | 292 | 56 | 50 | 106 | 14.02 |
| BaggingSVM | 320 | 221 | 74 | 121 | 195 | 26.49 |
| C-SVCF | 322 | 248 | 72 | 94 | 166 | 22.55 |

From the results in Table 2, it is shown that the lowest number of misclassified instances is AdaBoostSVM (14.02%), followed by C-SVCF, BaggingSVM, Bagging, and AdaBoost, respectively. As pointed out by Brodley and Friedl [4], eliminating 5 to 10% of outlier instances does not affect the improvement of classifier accuracy. The number of identifying and eliminating instances seems to make C-SVCF perform well in this instance. Following this, we demonstrate the effectiveness of outlier removal, using C-SVCF to compare with AdaBoost, Bagging, AdaBoostSVM and

BaggingSVM filtering, at 4 levels of outlier elimination including 5, 10, 15 and 20%, respectively.

## 4.2 Accuracy of Classification Results

In these experiments, the capability of the proposed filtering is compared with four filtering methods, including AdaBoost, Bagging, AdaBoostSVM and BaggingSVM. The effectiveness of the proposed filtering is measured using the average accuracy of seven classifiers involving C4.5, conjunctive rule, Naïve Bayes, NN-classifier, random committee, random forests and RBFNetwork, based on the models built from training sets. The overall performance of each classifier averages accuracy across all 10 partitions using 10-fold cross-validation. The experiment results are shown in Tables 3 and 4.

**Table 3.** The accuarcy of ensemble classifers at different levels of outlier removal

| Classifiers | 0% of Outlier Removal | 5% of Outlier Removal | | | 10% of Outlier Removal | | | 15% of Outlier Removal | | | 20% of Outlier Removal | | |
|---|---|---|---|---|---|---|---|---|---|---|---|---|---|
| | | A | B | SVM | A | B | SVM | A | B | SVM | A | B | SVM |
| C4.5 | 68.07 | 72.68 | 72.25 | 69.81 | 76.13 | 75.08 | 72.66 | 80.35 | 80.51 | 77.96 | 85.40 | 85.91 | 81.49 |
| Conjunctive Rule | 63.86 | 66.09 | 68.81 | 68.38 | 73.41 | 70.85 | 70.24 | 77.48 | 74.60 | 72.84 | 82.00 | 77.76 | 76.23 |
| Naïve Bayes | 68.21 | 72.39 | 71.67 | 70.82 | 75.23 | 74.92 | 74.77 | 79.55 | 78.12 | 77.16 | 84.38 | 82.17 | 81.15 |
| NN-Classifiers | 57.88 | 58.23 | 58.94 | 62.09 | 62.84 | 64.05 | 67.37 | 67.89 | 69.17 | 70.13 | 73.34 | 77.08 | 82.00 |
| Random Committee | 59.51 | 60.66 | 60.23 | 63.09 | 65.41 | 67.07 | 68.88 | 69.33 | 72.04 | 72.84 | 76.91 | 79.46 | 83.53 |
| Random Forests | 60.73 | 63.81 | 61.52 | 64.81 | 65.41 | 69.03 | 69.34 | 73.00 | 73.32 | 74.76 | 78.10 | 82.00 | 83.53 |
| RBF Network | 67.12 | 72.10 | 70.67 | 69.96 | 74.32 | 74.62 | 73.41 | 77.80 | 77.00 | 77.32 | 84.04 | 81.49 | 78.95 |
| Average | 63.63 | 66.57 | 66.30 | 66.99 | 70.39 | 70.80 | 70.95 | 75.06 | 74.97 | 74.72 | 80.60 | 80.84 | |

*Remark.* A refers to AdaBoost filtering and B refers to Bagging filtering.

**Table 4.** The accuarcy of SVM ensemble classifers at different levels of outlier removal

| Classifiers | 0% of Outlier Removal | 5% of Outlier Removal | | | 10% of Outlier Removal | | | 15% of Outlier Removal | | | 20% of Outlier Removal | | |
|---|---|---|---|---|---|---|---|---|---|---|---|---|---|
| | | A+S | B+S | SVM | A+S | B+S | SVM | A+S | B+S | SVM | A+S | B+S | SVM |
| C4.5 | 68.07 | 69.67 | 69.24 | 69.81 | 69.64 | 72.05 | 72.66 | - | 76.04 | 77.96 | - | 79.46 | 81.49 |
| Conjunctive Rule | 63.86 | 66.67 | 64.23 | 68.38 | 70.39 | 70.24 | 70.24 | - | 72.20 | 72.84 | - | 75.21 | 76.23 |
| Naïve Bayes | 68.21 | 70.39 | 70.96 | 70.82 | 73.11 | 73.41 | 74.77 | - | 76.04 | 77.16 | - | 79.46 | 81.15 |
| NN-Classifiers | 57.88 | 59.66 | 59.08 | 62.09 | 66.01 | 61.78 | 67.37 | - | 69.01 | 70.13 | - | 74.36 | 82.00 |
| Random Committee | 59.51 | 62.66 | 61.09 | 63.09 | 69.03 | 63.90 | 68.88 | - | 71.41 | 72.84 | - | 77.93 | 83.53 |
| Random Forests | 60.73 | 62.95 | 61.23 | 64.81 | 69.64 | 67.98 | 69.34 | - | 74.12 | 74.76 | - | 78.27 | 83.53 |
| RBF Network | 67.12 | 68.53 | 69.81 | 69.96 | 72.36 | 71.75 | 73.41 | - | 75.88 | 77.32 | - | 79.63 | 78.95 |
| Average | 63.63 | 65.79 | 65.09 | 66.99 | 70.03 | 68.73 | 70.95 | - | 73.53 | 74.72 | - | 77.76 | 80.98 |

*Remark.* A+S refers to AdaBoostSVM filtering; B+S refers to BaggingSVM filtering.

Tables 3 and 4 show that the outlier elimination methods using AdaBoost, Bagging AdaBoostSVM, BaggingSVM and C-SVCF improve the average prediction accuracy in seven classifiers including C4.5, conjunctive rule, Naïve Bayes, NN-classifier, random committee, random forests and RBFNetwork. This can be noticed, especially in the last row with average values. The average prediction accuracy is higher for all training sets with different levels of outlier elimination than in the original complete training set. Moreover, the results show that the C-SVCF filtering method is better than AdaBoost and Bagging in general. In addition, accuracies of each classifier have almost achieved 80% accuracy after removing 20% of outliers.

### 4.3 AUC Scores of Classification Results

In these experiments, the capability of the proposed filtering is compared with four of the aforementioned filtering methods. The effectiveness of the proposed filtering is measured using the average AUC score of seven of the abovementioned classifiers. The overall performance of each classifier averages the AUC scores across all 10 partitions using 10-fold cross-validation. The experiment results are shown in Tables 5 and 6.

**Table 5.** The AUC score of ensemble classifiers at different levels of outlier removal

| Classifiers | 0% of Outlier Removal | 5% of Outlier Removal | | | 10% of Outlier Removal | | | 15% of Outlier Removal | | | 20% of Outlier Removal | | |
|---|---|---|---|---|---|---|---|---|---|---|---|---|---|
| | | A | B | SVM | A | B | SVM | A | B | SVM | A | B | SVM |
| C4.5 | 69.67 | 73.50 | 70.00 | 72.60 | 75.80 | 76.30 | 74.90 | 79.80 | 80.90 | 79.90 | 85.30 | 85.60 | 80.40 |
| Conjunctive Rule | 63.86 | 67.40 | 69.50 | 64.70 | 70.30 | 68.50 | 71.30 | 74.80 | 72.60 | 69.70 | 80.50 | 73.90 | 72.90 |
| Naïve Bayes | 68.21 | 77.90 | 77.70 | 77.30 | 80.20 | 80.90 | 80.70 | 83.80 | 84.70 | 82.90 | 89.10 | 88.00 | 87.10 |
| NN-Classifiers | 57.88 | 57.70 | 58.60 | 61.70 | 62.10 | 63.80 | 67.00 | 66.90 | 68.80 | 69.80 | 72.20 | 76.90 | 81.60 |
| Random Committee | 59.51 | 60.60 | 60.20 | 65.80 | 65.60 | 69.50 | 70.40 | 71.40 | 74.20 | 76.70 | 78.60 | 81.70 | 89.20 |
| Random Forests | 60.73 | 66.00 | 66.20 | 68.90 | 69.50 | 74.20 | 75.00 | 75.20 | 78.60 | 81.50 | 82.80 | 86.90 | 90.80 |
| RBF Network | 67.12 | 76.90 | 76.00 | 76.10 | 78.80 | 80.10 | 78.80 | 81.80 | 83.80 | 82.50 | 89.00 | 87.70 | 86.90 |
| Average | 63.85 | 68.57 | 68.31 | | 71.76 | 73.33 | | 76.24 | 77.66 | 77.57 | 82.50 | 82.96 | |

*Remark.* A refers to AdaBoost filtering and B refers to Bagging filtering.

**Table 6.** The AUC score of SVM ensemble classifiers at different levels of outlier removal

| Classifiers | 0% of Outlier Removal | 5% of Outlier Removal | | | 10% of Outlier Removal | | | 15% of Outlier Removal | | | 20% of Outlier Removal | | |
|---|---|---|---|---|---|---|---|---|---|---|---|---|---|
| | | A+S | B+S | SVM | A+S | B+S | SVM | A+S | B+S | SVM | A+S | B+S | SVM |
| C4.5 | 69.67 | 70.70 | 70.20 | 72.60 | 76.00 | 72.80 | 74.90 | - | 75.30 | 79.90 | - | 80.60 | 80.40 |
| Conjunctive Rule | 63.86 | 68.90 | 66.40 | 64.70 | 68.60 | 68.90 | 71.30 | - | 68.40 | 69.70 | - | 72.20 | 72.90 |
| Naïve Bayes | 68.21 | 76.90 | 77.30 | 77.30 | 79.90 | 79.60 | 80.70 | - | 82.30 | 82.90 | - | 85.10 | 87.10 |
| NN-Classifiers | 57.88 | 59.20 | 58.70 | 61.70 | 65.70 | 61.50 | 67.00 | - | 68.20 | 69.80 | - | 73.70 | 81.60 |
| Random Committee | 59.51 | 64.10 | 62.50 | 65.80 | 75.60 | 65.60 | 70.40 | - | 75.20 | 76.70 | - | 81.60 | 89.20 |
| Random Forests | 60.73 | 67.40 | 65.40 | 68.90 | 75.90 | 71.50 | 75.00 | - | 78.80 | 81.50 | - | 85.00 | 90.80 |
| RBF Network | 67.12 | 75.10 | 75.40 | 76.10 | 79.40 | 78.50 | 78.80 | - | 80.10 | 82.50 | - | 85.20 | 86.90 |
| Average | 63.85 | 68.90 | 67.99 | *69.59* | 74.44 | 71.20 | *74.01* | - | 75.47 | 77.57 | - | 80.49 | *84.13* |

*Remark.* A+S refers to AdaBoostSVM filtering; B+S refers to BaggingSVM filtering.

Tables 5 and 6 show that the outlier elimination method using AdaBoost, Bagging AdaBoostSVM, BaggingSVM and C-SVCF, improve the average AUC scores in seven classifiers including C4.5, conjunctive rule, Naïve Bayes, NN-classifier, random committee, random forests and RBFNetwork. This can be noticed in the last row with average values. The results show that the AUC score of classifiers using C-SVCF filtering is better than AdaBoost, Bagging, AdaBoostSVM and BaggingSVM. However, eliminating the outlier up to 20% affects the significance improvement of model performance. Therefore, it could be concluded that filtering misclassified outliers using the C-SVCF method is most suitable for improving the overall performance of the models used in our dataset.

This paper has followed the hypothesis of most interest [3] [4] [27], which is to first filter the training set in pre-processing, and then uses classifiers to form the prediction models. Its advantage in identifying outliers allows us to show examples to experts who can distinguish outliers from normal instances. Moreover, outlier handling enables the learning algorithm to not only build models from good instances, but also get domain experts and users involved. Besides, we argue that with [4], even removal of 5 to 10% of outliers leads to significant performance improvement of learned models in the 7 classifiers of our framework. For example, the average accuracy of the prediction model have been improved 6.76%, 7.17%, 6.40%, 5.10% and 7.32%, after removing 10% of outliers using AdaBoost, Bagging, AdaBoostSVM, BaggingSVM and C-SVCF filtering, respectively. Similar to the average accuracy, the average AUC scores has been increased 7.91%, 9.48%, 10.59%, 7.35% and 10.16%, after removing 10% of outliers using AdaBoost, Bagging, AdaBoostSVM, BaggingSVM and C-SVCF filtering, respectively. In addition, filtering methods can lead to improvement of efficiency and scalability of classification models [28]. However, measuring efficiency and scalability of classification models after applying filtering methods are out of scope of this paper due to filtering methods affecting the efficiency and scalability of models generated from large and high dimension data sets.

## 5  Conclusion and Future Work

In this paper, a novel method of outlier removal has been proposed and applied to the task of breast cancer survivability analysis. The C-SVCF algorithm has been used to identify and eliminate the outliers from misclassified instances. Results show that outlier removal improves the results of instances selection by decreasing the insignificant outlier instances. This trend has been especially evidenced in the better result achieved by using C-SVCF in comparison with AdaBoost, Bagging and SVM ensembles. Moreover, the average prediction accuracy is improved 17.35% and the average AUC scores improved by 20.28% after removing 20% of outliers using C-SVCF filtering. However, for researchers it is hard to choose a suitable method for analysis intuitively, because one algorithm may perform well in one data set but perform badly in others. In future research, further investigations need to be conduced on developing and evaluating better classifiers for discovering better survivability models in our breast cancer data sets.

## References

1. Tsumoto, S.: Problems with Mining Medical Data. In: The Twenty-Fourth Annual International Conference on Computer Software and Applications, pp. 467–468 (2000)
2. Li, J., Fu, A.W.-C., He, H., Chen, J., Kelman, C.: Mining Risk Patterns in Medical Data. In: Proc. the Eleventh ACM SIGKDD International Conference on Knowledge Discovery in Data Mining, pp. 770–775 (2005)
3. Verbaeten, S., Assche, A.V.: Ensemble Methods for Noise Elimination in Classification Problems. In: Windeatt, T., Roli, F. (eds.) MCS 2003. LNCS, vol. 2709, pp. 317–325. Springer, Heidelberg (2003)
4. Brodley, C.E., Friedl, M.A.: Identifying and Eliminating Mislabeled Training Instances. J. Artificial Intelligence Research 1 (1996)
5. Brodley, C.E., Friedl, M.A.: Identifying Mislabeled Training Data. J. Artificial Intelligence Research. 11, 131–167 (1999)
6. John, G.H.: Robust Decision Trees: Removing Outliers from Databases. In: Proc. the First International Conference on Knowledge Discovery and Data Mining, pp. 174–179. AAAI Press, Menlo Park (1995)

7. Teng, C.M.: Applying Noise Handling Techniques to Genomic Data: A Case Study. In: Proc. the Third IEEE International Conference on Data Mining, p. 743 (2003)
8. Muhlenbach, F., Lallich, S., Zighed, D.A.: Identifying and Handling Mislabelled Instances. J. Intelligent Information Systems. 22(1), 89–109 (2004)
9. Hristovski, D., Peterlin, B., Mitchell, J.A., Humphrey, S.M.: Using Literature-Based Discovery to Identify Disease Candidate Genes. J. Medical Informatics. 74, 289–298 (2005)
10. Blanco, Á., Ricket, A.M., Martín-Merino, M.: Combining SVM Classifiers for Email Anti-Spam Filtering. In: Proc. the Ninth International Work-Conference on Artificial Neural Networks, pp. 903–910. Springer, Heidelberg (2007)
11. Han, J., Kamber, M.: Data Mining: Concepts and Techniques. Morgan Kaufmann, Elsevier Science, San Francisco (2006)
12. Vapnik, V.: Statistical Learning Theory. Wiley, New York (1998)
13. Yin, Z., Yin, P., Sun, F., Wu, H.: A Writer Recognition Approach Based on SVM. In: Multi Conference on Computational Engineering in Systems Applications, pp. 581–586 (2006)
14. Lallich, S., Muhlenbach, F., Zighed, D.A.: Improving Classification by Removing or Relabeling Mislabeled Instances. In: Proc. the Thirteen International Symposium on Foundations of Intelligent Systems, pp. 5–15 (2002)
15. Sun, J.-w., Zhao, F.-y., Wang, C.-j., Chen, S.-f.: Identifying and Correcting Mislabeled Training Instances. In: Future Generation Communication and Networking, pp. 244–250 (2007)
16. Xiao, Y., Khoshgoftaar, T.M., Seliya, N.: The Partitioning- and Rule-Based Filter for Noise Detection. In: Proc. IEEE International Conference on Information Reuse and Integration, pp. 205–210 (2005)
17. Yi, W., Fuyong, W.: Breast Cancer Diagnosis Via Support Vector Machines. In: Proc. the Twenty Fifth Chinese Control Conference, pp. 1853–1856 (2006)
18. Kohavi, R.: A Study of Cross-Validation and Bootstrap for Accuracy Estimation and Model Selection. In: Proc. the International Joint Conference on Artificial Intelligence, pp. 1137–1143 (1995)
19. Thongkam, J., Xu, G., Zhang, Y.: An Analysis of Data Selection Methods on Classifiers Accuracy Measures. J. Korn Ken University (2007)
20. Huang, J., Ling, C.X.: Using AUC and Accuracy in Evaluating Learning Algorithms. IEEE Transactions on Knowledge and Data Engineering, 299–310 (2005)
21. Hand, D.J., Till, J.R.: A Simple Generalisation of the Area under the ROC Curve for Multiple Class Classification Problems J. Machine Learning 45, 171–186 (2001)
22. He, X., Frey, E.C.: Three-Class ROC Analysis-the Equal Error Utility Assumption and the Optimality of Three-Class ROC Surface Using the Ideal Observer. IEEE Transactions on Medical Imaging, 979–986 (2006)
23. Woods, K., Bowyer, K.W.: Generating ROC Curves for Artificial Neural Networks. IEEE Transactions on Medical Imaging, 329–337 (1997)
24. Jiang, Y.: Uncertainty in the Output of Artificial Neural Networks. In: International Joint Conference on Neural Networks, pp. 2551–2556 (2007)
25. Witten, I.H., Frank, E.: Data Mining: Practical Machine Learning Tools and Techniques. Morgan Kaufmann, San Francisco (2005)
26. Chang, C.-C., Lin, C.-J.: Libsvm–a Library for Support Vector Machines, http://www.csie.ntu.edu.tw/~cjlin/libsvm
27. Gamberger, D., Šmuc, T., Marić, I.: Noise Detection and Elimination in Data Preprocessing Experiments in Medical Domains. J. Applied Artificial Intelligence. 14, 205–223 (2000)
28. Khoshgoftaar, T.M., Seliya, N., Gao, K.: Rule-Based Noise Detection for Software Measurement Data. In: Proc. IEEE International Conference on Information Reuse and Integration, pp. 302–307 (2004)

# An Empirical Study of Combined Classifiers for Knowledge Discovery on Medical Data Bases

Lucelene Lopes[1], Edson Emilio Scalabrin[1], and Paulo Fernandes[2]

[1] PPGTS, PUCPR, Curitiba, Brazil
`lucelene.lopes@pucpr.br`, `edson.scalabrin@ppgia.pucpr.br`
[2] PPGCC, PUCRS, Porto Alegre, Brazil
on sabbatical at LFCS, Univ. of Edinburgh,
Edinburgh, UK, CAPES grant 1341/07-3
`paulo.fernandes@pucrs.br`

**Abstract.** This paper compares the accuracy of combined classifiers in medical data bases to the same knowledge discovery techniques applied to generic data bases. Specifically, we apply Bagging and Boosting methods for 16 medical and 16 generic data bases and compare the accuracy results with a more traditional approach (C4.5 algorithm). Bagging and Boosting methods are applied using different numbers of classifiers and the accuracy is computed using a cross-validation technique. This paper main contribution resides in recommend the most accurate method and possible parameterization for medical data bases and an initial identification of some characteristics that make medical data bases different from generic ones.

## 1 Introduction

The need of data mining is an uncontestable reality in almost every domain of our very pervasive modern life. In the health domain, the possibility to extract information from medical data bases offer benefits going from the extraction of important insights about future policies to even automated diagnosis [5]. Despite of such enchanting motivation, not many studies pay attention to the particularity of the data mining problem in medical bases. Certainly, is not the absence of classification methods comparisons that prevents such study [7, 12].

Perhaps there are not many studies specific to medical data bases because it is not acknowledged if medical data bases actually differ from other (generic) data bases. The main motivation of this paper was forged under the assumption that there is such difference. In a practical point of view, this paper aims to contribute by the analysis of 16 medical and 16 non-medical (generic) data bases. We also apply three machine learning classification methods (C4.5 [11], Bagging [4] and Boosting [6]) observing the accuracy achieve by them over the 32 data bases.

This paper is organized as follows: Section 2 presents the medical and generic data bases to be studied in this paper discussing some of their characteristics. Section 3 presents a brief description of the machine learning classification methods C4.5, Bagging and Boosting. Section 4 presents the accuracy of each classification method applied to all data bases, and these results are discussed.

Y. Ishikawa et al. (Eds.): APWeb 2008 Workshops, LNCS 4977, pp. 110–121, 2008.

## 2   The Medical and Generic Data Bases

In this section, we present the data bases used in this paper. Table 1 present some details about these bases dividing them in two subsets. The first subset refers to bases with data extracted from medical information, and the second one refers to bases with data extracted from generic fields, *e.g.*, software development, games, wine recognition, biology, chemistry and physical phenomena.

The first column indicates the data base name and from which repository it was downloaded, as well as an identifier to each data base to be used further in this paper. The data bases marked with $^\triangle$ were obtained from the University of California Irvine repository [1], while data bases marked with $^\nabla$ were obtained from the University of West Virginia repository [3].

The second column contains the information about the data itself, namely: the number of instances in each data base (*instances*), the number of attributes excluding the class (*attributes*), the ratio between the number of attributes and the number of instances (*ratio*), and the percentile of missing values (*missing*) over the total (number of instances times number of attributes).

The last column contains the information about the classes of each data base, namely, the number of different classes (*classes*) and a rate indicating how unbalanced these classes are in the data base (*unbalance*). To have this rate independent of the number of instances of the base, the unbalance rate is computed as the ratio between the standard deviation of the number of instances in each class by a completely balanced distribution of instances among the classes, *i.e.*, the number of instances divided by the number of classes. Therefore, a data base with exactly the same number of instances in each class will have an unbalance rate equal to 0.00. A data base with a standard deviation as big as twice the number of instances by classes will have an unbalance rate equal to 2.00. For example, a data base with 300 instances and two classes, having 200 instances in a class and 100 in the other one, will have a unbalance rate equal to 0.47, which is computed as 0.7107 (the standard deviation) divided by 150 (the expected number of instances in each class in a perfect balanced situation[1].

An observation of Table 1 may lead to the conclusion that medical data bases are no different from generic ones. The missing information is perhaps the more noticeable difference between the subsets. All but one of the generic data bases (G11) have missing values and nearly two thirds of the medical bases (all but M06, M08, M09 and M15) have some missing values. The few number of instances compared to relatively large number of attributes is a even harder characteristic to assume, since while it is true for bases like M01, M02 and M07, the other medical data bases have a similar ratio as most of the generic data bases, and sometimes even considerably smaller than bases G06 and G16, not to mention the unique G11.

---

[1] A data base with 5 times this number of instances and the same unbalanced situation, *i.e.*, a data base with 1,500 instances with 1,000 from a class and 500 from the other, will also have the same unbalance rate of 0.47.

**Table 1.** Data Bases Characteristics

| Data Bases | id | Information Data | | | | Class Data | |
|---|---|---|---|---|---|---|---|
| | | attributes | instances | ratio | missing | classes | unbalance |
| Abalone$^\triangle$ | B01 | 8 | 4,177 | <0.01 | 0.00% | 28 | 1.41 |
| Arrythmia$^\triangle$ | B02 | 279 | 452 | 0.62 | 0.32% | 13 | 1.87 |
| Audiology$^\triangle$ | B03 | 69 | 226 | 0.31 | 2.03% | 24 | 1.58 |
| Balance$^\triangle$ | B04 | 4 | 625 | 0.01 | 0.00% | 3 | 0.66 |
| Breast cancer$^\triangle$ | B05 | 9 | 286 | 0.03 | 0.39% | 2 | 0.57 |
| Car Evaluation$^\triangle$ | B06 | 6 | 1,728 | <0.01 | 0.00% | 4 | 1.25 |
| CM1 software defect$^\triangledown$ | B07 | 21 | 498 | 0.04 | 0.00% | 2 | 1.14 |
| Datatrieve$^\triangledown$ | B08 | 8 | 130 | 0.06 | 0.00% | 2 | 1.17 |
| Desharnais$^\triangledown$ | B09 | 11 | 81 | 0.14 | 0.00% | 3 | 0.67 |
| Ecoli$^\triangle$ | B10 | 8 | 336 | 0.02 | 0.00% | 8 | 1.16 |
| Echo cardiogram$^\triangle$ | B11 | 11 | 132 | 0.08 | 5.10% | 3 | 0.40 |
| Glass$^\triangle$ | B12 | 10 | 214 | 0.05 | 0.00% | 6 | 0.83 |
| Heart(Cleveland)$^\triangle$ | B13 | 13 | 303 | 0.04 | 0.38% | 2 | 0.13 |
| Heart statlog$^\triangle$ | B14 | 13 | 270 | 0.05 | 0.00% | 2 | 0.16 |
| Hepatitis$^\triangle$ | B15 | 19 | 155 | 0.12 | 5.70% | 2 | 0.83 |
| JM1 software defect$^\triangledown$ | B16 | 21 | 10,885 | <0.01 | 0.00% | 2 | 0.87 |
| Kr-vs-kp$^\triangle$ | B17 | 36 | 3,196 | 0.01 | 0.00% | 2 | 0.06 |
| MW1 software defect$^\triangledown$ | B18 | 37 | 403 | 0.09 | 0.00% | 2 | 1.20 |
| Pima-diabetes$^\triangle$ | B19 | 8 | 768 | 0.01 | 0.00% | 2 | 0.43 |
| Post-operative$^\triangle$ | B20 | 8 | 90 | 0.09 | 0.42% | 3 | 1.05 |
| Primary-tumor$^\triangle$ | B21 | 17 | 339 | 0.05 | 3.92% | 21 | 1.18 |
| Reuse$^\triangledown$ | B21 | 27 | 24 | 1.13 | 0.93% | 2 | 0.35 |
| Solar Flare$^\triangle$ | B23 | 12 | 1,389 | 0.01 | 0.00% | 8 | 2.34 |
| Tic-Tac-Toe Endgame$^\triangle$ | B24 | 9 | 958 | 0.01 | 0.00% | 2 | 0.43 |
| Thyroid(Allhyper)$^\triangle$ | B25 | 29 | 2,800 | 0.01 | 5.61% | 4 | 1.93 |
| Thyroid(Hypothyroid)$^\triangle$ | B26 | 29 | 3,772 | 0.01 | 5.54% | 4 | 1.80 |
| Thyroid(Sick euthyroid)$^\triangle$ | B27 | 25 | 3,163 | 0.01 | 6.74% | 2 | 1.15 |
| Wbdc$^\triangle$ | B28 | 30 | 569 | 0.05 | 0.00% | 2 | 0.36 |
| Wisconsin breast cancer$^\triangle$ | B29 | 9 | 699 | 0.01 | 0.25% | 2 | 0.44 |
| Wine recognition$^\triangle$ | B30 | 13 | 178 | 0.07 | 0.00% | 3 | 0.19 |
| Yeast$^\triangle$ | B31 | 9 | 1,484 | 0.01 | 0.00% | 10 | 1.17 |
| Zoo$^\triangle$ | B32 | 17 | 101 | 0.17 | 0.00% | 7 | 0.89 |

In contradiction with the observation for these 32 data bases, spread ideas in the research and practionner communities usually claim that medical data bases are distinguishable due to large amounts of missing values, few instances, many attributes, *etc.*It is true, however, that the aspects presented in Table 1 do not encompass all the particularities of the data bases. The relationship between the instances, *e.g.*, the amount of contradiction or redundancy, may play an important role to distinguish medical from generic data bases. Even though the traditional ideas of the community are not clear noticeable, there must be special aspects concerning the medical data bases. As will be seem in Section 4, despite the lack of a clear distinction between the two subsets of data bases, the better performance of some classification methods to one group or another is noticeable.

# 3    The Knowledge Discovery Methods

There are several methods for knowledge discovery aiming classification, which is the most common application for medical data bases [5]. In this paper we focus on a classical approach represented by the C4.5 method [11] and two more refined approaches based on combining classifiers: Bagging [4] and Boosting [6]. Note that the methods Bagging and Boosting used in this paper also rely on C4.5 tree generation algorithm to obtain their combined classifiers. In this section we briefly describe these three methods.

## 3.1    C4.5 Method

The method C4.5 used in this paper is a java implementation of the algorithm proposed by Quinlan [11], sometimes referred as J48 [14]. The algorithm basic idea is to generate a decision tree based on a training set of instances. Each node of this tree, except the leaves, represents an attribute and the branches the choice of an attribute value. The leaves of the decision tree correspond to the possible classes.

The generation of the decision tree is made by choosing recursively an attribute to each node until the branches point clearly to one of the class. Therefore, this decision tree can be used as a classifier, *i.e.*, a new instance can be classified according to its attribute values.

The C4.5 method chooses attributes to the decision tree nodes by computing the entropy of subsets of instances, and choosing the atributes that give the greater information gain. Details on the algorithm itself and formulae for entropy and information gain can be found in [10, 11, 14].

According to the literature [4, 8], the C4.5 method is quite efficient for generating accurate classifiers for many data bases. However, as may be observed by the results in this paper, this method deals poorly with data bases where there is considerably high rate of missing attribute values. Unfortunately, this is frequently the case for medical data bases.

## 3.2    Bagging Method

Bagging was proposed by Breiman in 1996 [4] and its basic idea is to generate not only one, but a certain number of classifiers to a training data base. These classifiers are generated independently and the overall classification is made by a majority vote among the classifiers.

A Bagging method application that combines $k$ classifiers correspond to generate $k$ samples of the training set with $M$ instances. Each sample contains as many instances as the original training set $(M)$, but instead of using all instances of the original training set, the instances are uniformly sampled with repetition, *i.e.*, the sampling method picks $M$ times one instance of the original training set with an uniform distribution. Such sampling method will generate $k$ samples that randomly represent some aspects of the original data base.

To each sample, a classifier is generated independently, resulting in $k$ classifiers. The classification of a new instance will be performed applying each of these $k$ classifiers and the class chosen by the greater number of classifiers will be considered as the class of the new instance.

Making an analogy to real life decisions, the classification of new instances using the Bagging method can be viewed as the composition of a decision board where all voters have an opinion based on their randomly defined previous experiences (the random samples of the original training set).

According to the literature [4, 8], the Bagging method provides a better classification than one single C4.5 classifier generation (C4.5 method). This improvement in the accuracy is particularly noticeable for data bases with discrepancy among the instances, which is quite common to medical data bases.

## 3.3    Boosting Method

The Boosting application in this paper experiments is based on the Adaboost algorithm proposed by Freund and Schapire in 1996 [6]. Like Bagging, Boosting is also a method based on combining $k$ different classifiers. Actually the main differences between the methods Bagging and Boosting reside in the way samples are generated, and in the way the final classification is performed.

In the Bagging method the classifiers are generated according to samples chosen randomly, or let us say, independently from each other. The Boosting method uses a more refined way to sample the original training set, where the samples are chosen according to the accuracy of the previously generated classifiers. In fact, each of the $k$ steps of classifier generation takes into account the accuracy of the classifiers generated in the previous steps, this accuracy is usually describe as $\alpha^{(i)}$ for the $i^{th}$ classifier generated. The Boosting method defines a vector with a weight for all instances of the original training set (a vector with $M$ positions) and initializes this vector with an equiprobable distribution. The first sample is generated according to this equiprobable weight vector, *i.e.*, it samples the original training set just like the Bagging method. After each classifier accuracy tested by applying it to the whole original training set, each weight corresponding to correctly classified instances is multiplied by $\alpha^{(i)}$, while uncorrected classified instances' weight is divided by $\alpha^{(i)}$. Therefore, incorrectly classified instances will be more likely to be included in the next sample.

The second difference from Bagging is the way Boosting computes the votes of the classifiers when actually classifying a new instance. While in Bagging a simple majority vote is assumed, in Boosting each classifier will weight his vote with its own accuracy $(\alpha^{(i)})$.

Once again, making an analogy with real life decisions, Boosting classification is similar to a board decision, but unlike Bagging which takes random components to compose the board, Boosting composes its board with specialists with rather different points of views. In order to vote, the opinions of these specialists are weighted according to how relevant their specialty is according to the training set.

According to the literature [4, 7], Boosting method is quite efficient to classify noise-free data set, *i.e.*, data sets where there is no discrepancy in the data, there is no missing values, *etc.*. However, such conclusions are made, at the authors' best knowledge, based on empirical experimentations only.

## 4   The Experiments

The experiments conducted on this paper use the *WEKA - Waikato Environment for Knowledge Analysis*, version 3.4.8 [14]. The Bagging and Boosting methods were applied using from 10 to 50 classifiers. The data sets were tested using a cross-validation technique [10] that basically combines different parts of the data set as training set and testing set. Specifically, for this paper we use a cross-validation technique that randomly chooses 10 disjoint 10% of the original data set sized samples and runs the classification to each of the 10 samples using the remaining 90% of the original data set as training set and the sample itself as testing set. The accuracy results expressed correspond to the average accuracy of each 10 classification runs. Such method of experiment consists in use the same data base as training and testing set. The random choice of samples, therefore, plays a role in the accuracy and all tests should take this into account. For the experiments in this paper a fixed random seed was used.

Table 2 presents the percentile of correctly classified instances (accuracy) achieved for all data bases in Table 1 applying the three tested methods. For the C4.5 method there is only one accuracy value, since this method does not combine classifiers. For methods Bagging and Boosting the results present the values achieved for 10, 20, 30, 40 and 50 classifiers. The last column (*best accuracy*) indicates the best method, which is assumed to be the method with the higher accuracy. This value is also indicated by the bold face accuracy in each row. Observing Table 2 we notice that the difference between the accuracy of methods applied to a same data base is usually quite small, specially for method with high accuracy, *i.e.*, all results over 90%. Only particular cases present significant changes, *i.e.*, a difference larger than 10% points from the worst to the best accuracy, *e.g.*, 14% in M10, 13% in G13, and 12% in G02.

At this point it is important to keep in mind what the accuracy results mean. As said before, our experiments are executed over data bases using a cross-validation technique. It implies that the same data base is used as training set and testing set and therefore a random choice of samples affects the results. Particularly for small data bases, like M10, G06 and G11, it may result in some imprecisions of an order of 1%.

The numerical differences are even smaller when observing the changes of accuracy due to an increment of the number of classifiers. In fact, only Boosting accuracy had a significant change due to the increase of the number classifiers, *e.g.*, 3.12% for G06, 2.70% for M01, 2.56% for G13, and 2.02% for M09. Notice that these improvements in the accuracy for Boosting happen mostly on data bases where Bagging was a better choice, since among these bases, only G13 has Boosting as the most accurate method.

### 4.1   Individual Analysis

Observing some particular cases of the accuracy, Figure 1 shows the results for data base M02 and M16. Those two results demonstrate different success cases of the Boosting for medical bases. For base M02 neither of the combined

**Table 2.** Accuracy of Methods C4.5, Bagging and Boosting for All Data Bases

| data bases | C4.5 | Bagging 10 clf. | 20 clf. | 30 clf. | 40 clf. | 50 clf. | Boosting 10 clf. | 20 clf. | 30 clf. | 40 clf. | 50 clf. | best accuracy |
|---|---|---|---|---|---|---|---|---|---|---|---|---|
| M01 | 65.78 | 73.17 | 74.45 | 74.53 | 74.89 | **74.91** | 71.62 | 73.04 | 73.77 | 73.92 | 74.32 | Bagging |
| M02 | 77.26 | 80.84 | 80.62 | 80.62 | 80.75 | 80.88 | **84.75** | 84.61 | 84.39 | 84.66 | 84.61 | Boosting |
| M03 | **74.28** | 72.71 | 72.67 | 72.57 | 73.06 | 73.13 | 66.89 | 67.07 | 66.57 | 66.43 | 66.43 | C4.5 |
| M04 | 57.09 | 61.12 | 61.40 | 61.55 | 61.95 | **62.53** | 60.64 | 61.25 | 60.35 | 61.98 | 61.46 | Bagging |
| M05 | 77.13 | 79.02 | 79.73 | 79.80 | 79.47 | 80.10 | 78.59 | **80.21** | 80.11 | 80.18 | 80.11 | Boosting |
| M06 | 78.15 | 80.59 | **81.22** | 81.07 | 81.15 | 81.19 | 78.59 | 79.78 | 79.74 | 80.41 | 80.44 | Bagging |
| M07 | 79.22 | 80.73 | 81.04 | 81.17 | 81.37 | 81.50 | 82.38 | 82.79 | 83.65 | 83.05 | **84.15** | Boosting |
| M08 | 99.44 | 99.42 | 99.45 | 99.44 | 99.44 | 99.46 | 99.59 | 99.61 | **99.62** | 99.61 | 99.60 | Boosting |
| M09 | 74.49 | 75.65 | 76.21 | **76.27** | 76.21 | 76.05 | 71.69 | 72.61 | 73.07 | 73.36 | 73.71 | Bagging |
| M10 | **69.67** | 68.67 | 68.89 | 68.56 | 68.89 | 69,00 | 55.33 | 56.33 | 56.33 | 56.33 | 56.33 | C4.5 |
| M11 | 41.39 | 43.90 | 44.93 | 44.64 | 44.57 | **44.96** | 41.65 | 41.65 | 41.65 | 41.65 | 41.65 | Bagging |
| M12 | 99.54 | 99.58 | 99.57 | 99.59 | 99.59 | 99.59 | 99.65 | 99.32 | **99.69** | 99.68 | 99.69 | Boosting |
| M13 | 98.54 | 98.56 | 98.61 | 98.61 | 98.63 | 98.63 | 98.61 | 98.65 | **98.67** | 98.64 | 98.65 | Boosting |
| M14 | 97.94 | 97.94 | 97.93 | 97.94 | **97.95** | 97.94 | 97.56 | 97.76 | 97.76 | 97.80 | 97.85 | Bagging |
| M15 | 93.27 | 95.15 | 95.49 | 95.29 | 95.54 | 95.55 | 96.05 | 96.64 | **96.91** | 96.82 | 96.87 | Boosting |
| M16 | 95.01 | 96.07 | 96.12 | 96.24 | 96.17 | 96.24 | 96.08 | 96.38 | 96.47 | 96.51 | **96.61** | Boosting |
| G01 | 20.99 | 23.37 | 23.69 | **23.84** | 23.84 | 23.84 | 21.87 | 22.30 | 22.34 | 23.35 | 23.35 | Bagging |
| G02 | 32.39 | 19.94 | 20.61 | 20.21 | 20.21 | 20.00 | **32.39** | 32.39 | 32.39 | 32.39 | 32.39 | Boosting |
| G03 | 92.22 | 93.42 | 93.51 | 93.62 | 93.54 | 93.59 | 95.85 | 96.49 | 96.66 | 96.66 | **96.72** | Boosting |
| G04 | 88.05 | 88.84 | 88.80 | 89.08 | 89.10 | **89.18** | 87.33 | 87.43 | 87.39 | 87.49 | 87.51 | Bagging |
| G05 | 90.08 | 90.08 | 90.38 | 90.69 | **90.77** | 90.38 | 87.62 | 87.46 | 87.62 | 87.85 | 87.62 | Bagging |
| G06 | 66.92 | 70.25 | 71.33 | **72.07** | 71.22 | 71.71 | 68.44 | 69.83 | 71.04 | 71.56 | 71.08 | Bagging |
| G07 | 82.83 | 84.03 | 83.87 | 84.02 | 84.16 | **84.31** | 76.23 | 75.07 | 74.93 | 74.83 | 75.13 | Bagging |
| G08 | 97.33 | 97.57 | 97.52 | 97.61 | 97.57 | **97.61** | 97.33 | 97.33 | 97.33 | 97.33 | 97.33 | Bagging |
| G09 | 79.73 | 81.29 | 81.71 | 81.84 | 81.87 | **81.94** | 79.12 | 79.17 | 79.17 | 79.17 | 79.17 | Bagging |
| G10 | 91.42 | 91.64 | 91.82 | **91.86** | 91.86 | 91.86 | 90.00 | 90.20 | 90.20 | 90.20 | 90.20 | Bagging |
| G11 | 95.17 | 93.83 | 94.67 | 94.83 | 94.83 | 94.83 | 94.00 | **95.33** | 95.33 | 95.33 | 95.33 | Boosting |
| G12 | 99.14 | **99.14** | 99.14 | 99.14 | 99.14 | 99.14 | 98.58 | 98.59 | 98.61 | 98.60 | 98.62 | Bagging |
| G13 | 85.57 | 92.99 | 94.13 | 94.55 | 94.44 | 94.73 | 96.36 | 98.17 | 98.58 | 98.75 | **98.92** | Boosting |
| G14 | 93.82 | 94.94 | 94.94 | 96.07 | 96.07 | 96.07 | 96.63 | **97.19** | 97.19 | 97.19 | 97.19 | Boosting |
| G15 | 50.38 | 52.30 | 52.42 | **52.43** | 52.41 | 52.40 | 50.32 | 50.32 | 50.32 | 50.32 | 50.32 | Bagging |
| G16 | 92.61 | 93.41 | 93.21 | 93.30 | 93.30 | 93.20 | **97.35** | 97.35 | 97.35 | 97.35 | 97.35 | Boosting |

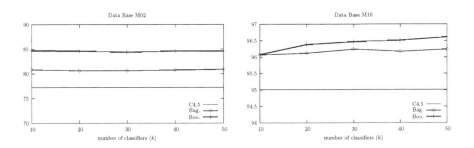

**Fig. 1.** Accuracy evolution for bases M02 and M16

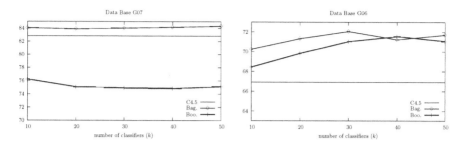

**Fig. 2.** Accuracy evolution for bases G07 and G06

classifiers methods presented an evolution as the number of classifiers grown, but Boosting accuracy was clearly above Bagging and C4.5 values. For base M16 both methods presented some evolution increasing the number of classifiers, but Boosting accuracy gains were bigger than Bagging's. It is also noticeable that even being both from the medical subset, those two data bases are quite different. While M02 has a large number of attributes compared to the number of instances (ratio 0.31), M16 has a much larger number of instances (ratio 0.01). Also, M02 is one of the most unbalanced bases (rate 1.58) with very distinct number of instances in each of its 24 different classes. M16 is rather balanced data with 458 instances in class *benign* and 241 instances in class *malign* (rate 0.44). It does look like the only similarity justifying that both bases have a better accuracy using Boosting is the simple fact that they are both medical data bases.

Observing now successful cases of Bagging in generic bases, Figure 2 shows the accuracy results for bases G07 and G06. G07 is a typical success case of Bagging method which gave results superior to those of C4.5 and Boosting method is simply disastrous. For G06, on the contrary, Bagging and Boosting gave good results, but here the number of classifiers was decisive to choose Bagging as the more accurate method. Both methods increase the accuracy until 30 classifiers, where Bagging reaches its peak. However, after that number of classifiers Bagging accuracy becomes slightly lower and Boosting results become similar and occasionally higher.

Unlike most of the previous results, it was surprising the better accuracy of method C4.5 for M03 and the results for G14. In both cases (Figure 3), the stagnation of Bagging method was not expected, because of Bagging robustness [9, 13]. For M03, like for G07, Boosting had a clearly inferior accuracy, but the surprise came from the fact that Bagging could not surpass C4.5 accuracy, since it practically did not improve accuracy with the addition of classifiers. G14 results present a similar behavior of Bagging accuracy, but in this case it was Boosting that offer the best accuracy, while C4.5 was clearly inferior.

### 4.2 Overall Analysis

Concerning the method that presents the best accuracy, Bagging was the best one for 16 bases, while Boosting was for 14 bases, and C4.5 was just for 2 bases.

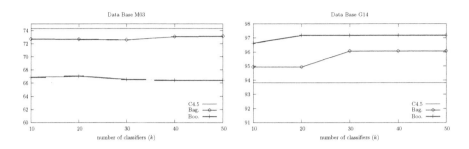

**Fig. 3.** Accuracy evolution for bases M03 and G14

However, observing separately the medical and generic databases we notice a different distribution, as shown in Figure 4. Boosting seems to be more adequate to medical data bases, being the most accurate in 8 of the 16 medical data bases. This result is somewhat surprising because it shows that Boosting method, usually said to be more accurate for noise-free data bases [7], has a better performance for medical data where one might expect more noise on the information [5]. Also surprising is the fact that the simpler C4.5 method, which usually gives considerably stiffer classification than the other methods, was more accurate for 2 medical data bases.

Trying to look closer to the data bases where each method had the best accuracy, we did not found any particular similarity for the characteristic as stated in Table 1. Table 3 presents the analysis of the bases grouped according to the most accurate method. The first group shows the analysis of minimum, average and maximum values for the 16 bases where Bagging was the most accurate method. Analogously, the second and third groups show the same values for the 14 bases where Boosting was more effective and the 2 bases where C4.5 gave the best results, respectively.

A definitive conclusion can not arise neither observing the average values nor the intervals between the minimum and maximum values. Considering only the average values no clear order among Bagging, Boosting and C4.5 values can be established. The intervals comparison is of very little help as well, since

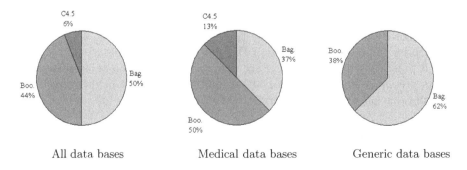

**Fig. 4.** Most Accurate Method Distribution

**Table 3.** Data Bases Characteristics Grouped by the Most Accurate Method

| M01, M04, M06, M09, M11, M14, G01, G04, G05, G06, G07, G08, G09, G10, G12, G15 | | | | | | |
|---|---|---|---|---|---|---|
| **Bagging** *attributes* | *instances* | *ratio* | *missing* | *classes* | | *unbalance* |
| min. | 8.0 | 81 | <0.00 | 0.00% | 2.0 | 0.16 |
| avg. | 31.1 | 1,545 | 0.08 | 1.01% | 7.1 | 1.07 |
| max. | 279.0 | 10,885 | 0.62 | 6.74% | 28.0 | 2.34 |

| M02, M05, M07, M08, M12, M13, M15, M16, G02, G03, G11, G13, G14, G16 | | | | | | |
|---|---|---|---|---|---|---|
| **Boosting** *attributes* | *instances* | *ratio* | *missing* | *classes* | | *unbalance* |
| min. | 4.0 | 24 | <0.00 | 0.00% | 2.0 | 0.06 |
| avg. | 22.1 | 1,095 | 0.14 | 1.46% | 4.5 | 0.78 |
| max. | 69.0 | 3,772 | 1.13 | 5.70% | 24.0 | 1.93 |

| M03, M10 | | | | | | |
|---|---|---|---|---|---|---|
| **C4.5** *attributes* | *instances* | *ratio* | *missing* | *classes* | | *unbalance* |
| min. | 8.0 | 90 | 0.03 | 0.39% | 2.0 | 0.57 |
| avg. | 8.5 | 188 | 0.06 | 0.41% | 2.5 | 0.81 |
| max. | 9.0 | 286 | 0.09 | 0.42% | 3.0 | 1.05 |

there are large overlappings preventing any clear rule to choose either of the methods based on their characteristics. That being said, it cannot be ignored that something different occurs with the medical data bases, since the robust Bagging method, so accurate in the generic bases, was not so effective in the 16 medical bases.

## 5   Conclusion

The conclusion of such empirical comparison of accuracy results for medical and generic data bases is a rather difficult task. The analyzed 32 data bases are public data that had been used in many research efforts to increase both the understanding of the classification methods, and the identification of the bases particularities. Such research efforts are far from easy, after all, in the vast majority of the cases, the data sets are sufficiently large (sometimes really huge) to justify a machine aided data mining approach. Therefore, the study of the particularities of such data bases is by definition a problem too complex to be handled manually. In many ways, the search for the reasons why a machine learning method does (or doesn't) work is a meta-problem, and, naturally, it leaves the door open for empirical studies.

In the present work, we try to analyze this problem with an approach starting from the question: *what makes a medical data base different from others?* Considering the aspects observed in Section 2, this answer cannot be answer yet, since there is no clear distinction. Nevertheless, the accuracy results point in different direction, since Boosting method, usually not as robust as Bagging, had a better overall accuracy for medical data bases. In fact, it is very likely that yet unknown characteristics of medical data bases play a major role in the adequacy of Boosting method. Maybe a deeper analysis of the data inside the bases may offer insights. Perhaps there is some similarities between the instances, perhaps the contradictions in medical data behaves according to a particular pattern. The fact is that this research is an open subject.

Despite this ignorance of the reasons, it arises as a contribution from the present work the empirical conclusion that Boosting method seems to be the best bet to medical data mining. It is also recommendable to not waste too much time thinking about the number of classifiers, since the rare accuracy improvements were very small. Even thou these Boosting recommendations could be numerically concluded, in some cases classification accuracy was better served by Bagging, or the simpler C4.5. Once again, the lack of explanations why the medical data bases behave like that leaves room for doubts in choosing the classification method.

The future works for the study presented here are the natural pursuit of the reasons why Boosting generally presents a better accuracy for medical data bases. Another interesting line of work is to study the impact of attribute selection in medical data bases. The benefits of attribute selection in generic data bases have been discussed in the area for long time, however the results presented here could also suggest that the behavior of such technique to medical data bases may present is not necessarily the same.

# References

1. Asuncion, A., Newman, D.J.: UCI Machine Learning Repository. Repository, Irvine, CA: University of California, Department of Information and Computer Science (2007), http://www.ics.uci.edu/~mlearn/MLRepository.html
2. Bauer, E., Kohavi, R.: An empirical comparison of voting classification algorithms: Bagging, Boosting and variants. Machine Learning 36(1/2), 105–139 (1999)
3. Boetticher, G., Menzies, T., Ostrand, T.: PROMISE Repository of empirical software engineering data. Repository, West Virginia University, Department of Computer Science (2007), http://promisedata.org/
4. Breiman, L.: Bagging predictors. Machine Learning 24(2), 123–140 (1996)
5. Cios, K.J.: Medical Data Mining and Knowledge Discovery. Studies in Fuzziness and Soft Computing, vol. 60. Springer, Heidelberg (2001)
6. Freund, Y., Schapire, R.E.: Experiments with a new boosting algorithm. In: Int. Conf. on Machine Learning, pp. 148–156 (1996)
7. Kotsianti, S.B., Kanellopoulos, D.: Combining Bagging, Boosting and Dagging for classifications problems. In: Apolloni, B., Howlett, R.J., Jain, L. (eds.) KES 2007, Part II. LNCS (LNAI), vol. 4693. Springer, Heidelberg (2007)
8. Li, J., Cercone, N.: Assigning Missing Attribute Values Based on Rough Sets Theory. In: IEEE Int. Conf. on Granular Computing. IEEE Computer Society Press, Los Alamitos (2006)
9. Melville, P., Mooney, R.: Constructing Diverse Classifer Ensembles using Artificial Training Examples. In: Proceedings of IJCAI 2003, Acapulco, Mexico, pp. 505–510 (2003)

10. Mitchell, T.: Machine Learning. McGraw-Hill, New York (1997)
11. Quinlan, J.: Induction of decision trees. Machine Learning 1(1), 81–106 (1986)
12. Quinlan, J.: Bagging, Boosting and C4.5. In: Proceedings of AAAI/IAAI. The MIT Press, Cambridge (1996)
13. Schapire, R.E., Freund, Y., Bartlett, P., Lee, W.S.: Boosting the margin: A New explanationfor the effectiveness of voting methods. The Annals of Statistics 26, 1651–1686 (1998)
14. Witten, I.H., Frank, E.: Data mining: practical machine learning tools and techniques, 2nd edn. Morgan Kaufmann, San Francisco (2005)

# Tracing the Application of Clinical Guidelines*

Eladio Domínguez, Beatriz Pérez, and María A. Zapata

Dpto. de Informática e Ingeniería de Sistemas,
Facultad de Ciencias, Edificio de Matemáticas,
Universidad de Zaragoza, 50009 Zaragoza, Spain
{noesis,beaperez,mazapata}@unizar.es

**Abstract.** Clinical guidelines have been developed with different aims and interests related to healthcare quality improvement. The development of storage mechanisms concerning the application of guidelines can provide significant benefits to healthcare.

In this paper, we present a Model Driven Development (MDD) based approach for the automatic generation of storage structures for recording the information generated during the application of clinical guidelines. The work presented is part of a larger project which aims at developing decision support systems for the application of guidelines. Our approach is illustrated with a Spanish guideline based on a guideline published by the National Guidelines Clearing House (NGC).

**Keywords:** Clinical guidelines, Health data storage, Decision Support.

## 1   Introduction

In healthcare daily practice, it is widely accepted that documentation constitutes a key element for demonstrating that care is as expected [18]. There are many works in the literature which have emphasized the importance of documenting every intervention or patient instruction whatever the patient's condition [18,21]. On the other hand, clinical guidelines are developed with different aims and interests concerning healthcare quality, including the improvement of clinical practice, support to medical decision-making or reduction of inappropriate variations in everyday clinical practice [24,29].

Based on the recognized importance of documentation in the healthcare context, we have focused our attention on the particular case of developing mechanisms for recording the history of the application of clinical guidelines. In particular, we are studying, as a long term research goal, the development of ubiquitous decision support systems (UDSS) with trace recording capabilities, for the application of clinical guidelines [30,31]. We propose that such a system implements guidelines in such a way that it guides the physician without allowing her decisions to diverge from the strategy defined in the guideline. In this

---

* This work has been partially funded by the Spanish Ministry of Education and Science, project TIN2005-05534 and the FPU grant AP2003-2713, and by the Ministry of Industry, Tourism and Commerce, project SPOCS (FIT-340001-2007-4).

Y. Ishikawa et al. (Eds.): APWeb 2008 Workshops, LNCS 4977, pp. 122–133, 2008.

manner, we ensure that the care provided to a patient is in compliance with guideline recommendations. As for the trace recording capabilities, the UDSS would store the information generated during the application of the guideline (such as physician's decisions or patient's states). We consider that the storage of this information can contribute to the improvement of several healthcare aspects, such as the quality of patient care or the prevention of medical errors.

In order to obtain the UDSS for each guideline automatically, we propose to follow a Model Driven Development [26,33] (MDD) approach. By using this MDD approach, we ensure that the development of the system is carried out in an efficient and cost–effective way. In particular, we represent the dynamics of each clinical guideline by using a UML statechart [27]. When a system is developed within a MDD context, the history of the system execution can be expressed as the trace of its model execution [16]. In our specific case, given the statechart, a database schema for storing the trace of the guideline application is automatically generated. This process is carried out by following a set of successive schema mappings [15] that we have implemented in a MDD setting using two different MDA–based tools (ATL [5] and MOFscript [20]).

We proceed as follows: In Section 2 we present an overview of the development of UDSSs for clinical guidelines. We describe the definition of the conceptual database schema from a statechart representing a guideline in Section 3. Section 4 describes the generation of the database from the conceptual model by using a set of schema mappings. The implementation of our proposal is presented in Section 5. In Section 6 we discuss related work. Finally, our conclusions and future work are set out in Section 7.

## 2   Overview

Recently we have proposed a MDA–based approach to automatically develop ubiquitous decision support systems (UDSS) for clinical guidelines [30,31]. Among the goals each UDSS should achieve, we want to highlight the following: (1) that it guides the physician during the application of the guideline in a very specific way in order to help in her decision making and (2) that it automatically records the history of the application of the guideline to patients. Then, information such as patients' conditions, physicians' decisions, treatments and clinical tests applied to patients, will be recorded for different purposes. From now on, we use the term *trace* to refer to the history of the application of a guideline to a patient. In addition, we define a *trace database* as the persistent database which stores those trace records.

The storage of the information concerning the application of a guideline by using the UDSS can provide several significant benefits to the healthcare environment. Many works have claimed that thorough documentation of different aspects of patient care could be key evidence in a medical negligence case (see e.g. [18]). We consider that the traces recorded by using the UDSS could also be used to reduce malpractice litigation. Furthermore, such traces can be used as a reference in future encounters with the patient when the same guideline is reapplied [12]. In addition, the regular and necessary revision and update of clinical guidelines [32] can be improved by means of documenting. The traces recorded by using our UDSS can be considered as a resource for ongoing changes in the definition of guidelines over time.

The architecture of the UDSS for a clinical guideline is presented in Fig. 1. Following our approach presented in [9,30,31], the dynamics of each guideline is represented by using a UML statechart. In particular, we have defined, taking into account the specialities of the clinical data, a set of representation patterns in order to assist in the modelization process [31]. From this model, the main modules which constitute the UDSS for the guideline are automatically generated by following a MDA–approach. In the bottom left part of Fig. 1 we depict two modules that form part of the UDSS. Firstly, the "execution module" corresponds with the Java implementation of the statechart and allows the physician to apply the UDSS to patients. Secondly, the "platform library" provides standard services of the system related to the implementation of the presentation layer and the data layer. In this paper, we propose to include in the UDSS a new component (trace database) that allows the persistence of the guideline application (see the bottom right part of Fig. 1). The instances of the database come from the execution of the Java program that constitutes the "execution module" of each UDSS. The way in which the generated instances are recorded in the trace database remains an issue for future work.

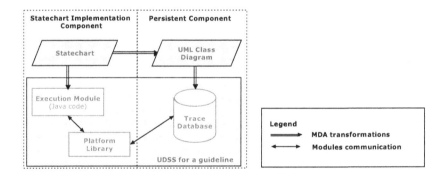

**Fig. 1.** The architecture of the UDSS for a guideline

In order to design and develop the trace database, we propose the use of a MDD approach. Considering that the information represented in the statechart modeling a guideline is the data intended to be recorded in the database, we use the statechart as the basis for creating the database schema. Then, as is depicted in Fig. 1, from each statechart we propose to define a UML class diagram as the conceptual model of the trace database. Based on this class diagram, we propose to carry out a set of schema mappings to obtain the database schema.

**Case Study.** In order that the reader can have a better understanding of our approach, we will use as a case study a Spanish guideline used for the management of infections related to intravenous catheters. This guideline is based on a guideline published by the US National Guidelines Clearing House (NGC) [1]. In particular, we will use a part of the statechart (see Fig. 2(a)) that represents the dynamics of this guideline. This is a UML orthogonal state that models the patient's behavior during the performance of two clinical tests, the Maki and Hemoccult tests. Regarding the Maki test (see left part of Fig. 2(a)), firstly the patient is waiting for the catheter (CVC) removal. When the catheter is removed, then the Maki test is ordered and the patient changes from that first state to that of waiting for the Maki test results. When the results finally arrive, they are recorded. With regard to the Hemoccult test (see right part of Fig. 2(a)), the process is similar. The class diagram defined for this orthogonal state is depicted in Fig. 2(b) and will be explained in the next section.

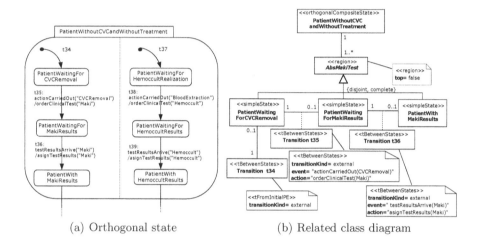

(a) Orthogonal state                    (b) Related class diagram

**Fig. 2.** Part of the statechart for a specific guideline and the related class diagram

## 3   Conceptual Representation of the Trace Database

In this section, we explain our approach to the conceptual modeling of the trace database from a statechart representing a clinical guideline. It is widely recognized that database systems are best designed first at a conceptual level. Then, the resulting conceptual model is used as the base for the subsequent development phases. Several approaches have addressed the use of UML class diagrams to express conceptual database models [2,8]. As is claimed in [8], class diagrams, unlike other conceptual data models like Entity Relationship diagrams (ERD), provide a more object-oriented approach to database design, supporting the specification of database structure and behavior.

Taking this into account, from each statechart, we propose to define a UML class diagram as the conceptual model of the trace database. To our knowledge, there is no criteria to follow for creating a class diagram from a statechart. Our proposal is based on the definition of (1) a UML profile [27] called the

UML Profile for *statechart execution persistence* (SEP profile) and (2) a set of transformation patterns to assist in the transformation process.

**The SEP Profile.** Since we represent each guideline by a statechart, the way of storing the information generated during the application of that guideline is strictly related to the execution of the statechart. On the other hand, the execution of the statechart depends on the semantics of the elements of the statechart. Therefore, we consider it necessary to represent in the UML class diagram the specific semantics of each element which constitutes the statechart. Nevertheless, it is not possible to represent this specific information with the native elements of UML class diagrams. In these cases, UML profiles [27] are defined to provide extension mechanisms to the UML standard. For this reason, we have created the SEP profile, so that some elements of the UML statechart metamodel have stereotypes of the UML profile associated to them.

The main aim of this profile is to give, by using its stereotypes, a simple mechanism to identify, for each class in the class diagram, the type of UML statechart element to which it corresponds. Then, these stereotypes will be implemented as *triggers* in the trace database. Each of these triggers will allow us to control the way in which the information is recorded by following the semantics of the UML statechart element to which each trigger corresponds.

For example, the semantics of UML orthogonal states indicates that whenever an orthogonal state is entered, each one of its orthogonal regions is also entered [27]. Taking this into account, we have defined the «OrthogonalCompositeState» stereotype to be applied to each class in the class diagram which comes from an orthogonal state. When this stereotype is implemented in the database, a trigger is created in order to ensure that the information associated with the entry of the object to all orthogonal regions is recorded in the database.

We have defined four stereotypes: «Region», «State», «TransitionBetweenStates» and «TransitionNotBetweenStates», where the three latter are specialized into several stereotypes. The UML elements extended by the stereotypes of the profile are the UML class and association class elements. We do not include more details about the defined profile for space reasons.

**Transformation Patterns.** The underlying idea of the transformation patterns is based on some ideas proposed in [23]. In that paper, the authors provide an approach for generating executable code from UML statechart diagrams in the Java object-oriented language, by extending the state design pattern [13]. These authors use UML class diagrams as the conceptual model of the Java program that implements the statechart. We have extended this approach by adding several changes (among them, the definition and application of the SEP profile) to adapt it to our specific purpose, using the class diagram as the conceptual model of the trace database. The main idea of our transformation patterns is that: (1) each state in the statechart is transformed into a UML class with a specific stereotype of the SEP profile depending on the type of state (simple, orthogonal or non orthogonal); (2) each transition is represented as either a UML class or a UML association class, depending on the type of its source and target vertex; also, the type of source and target vertex is used to apply the corresponding stereotype to the class (or association class); (3) each region is translated into an abstract UML class with the «region» stereotype.

For example, a part of the class diagram defined for the orthogonal state of Fig. 2(a) is shown in Fig. 2(b). In particular, the region related to the performance of the Maki test is represented by an abstract class called *AbsMakiTest* stereotyped by «region». This class is the parent class of the specific classes which correspond to the translation of the states contained in the region (*Patient WaitingForCVCRemoval*, *PatientWaitingForMakiResults* and *PatientWith-MakiResults*). These specific classes are stereotyped by «simpleState», since they correspond to the translation of simple states in the orthogonal state. Additionally, stereotyped classes associated to transitions are created. In this particular context, we record the information relating to the patients' specific states during the application of the guideline, represented by the statechart.

In addition, we want to record in the trace database temporal data which is not directly related to the statechart. Since the application of guidelines generates temporal data, we have defined the trace database as a temporal database. Of the four kinds of database (Snapshot, Historical, Roll–back and Bi–temporal) that can be defined considering the temporal–dimension taxonomy (transaction time, valid time and user–defined time) [11], we have decided to define ours as a bi–temporal database (we consider both transaction and valid time). We have based our decision on the fact that these databases have the ability to answer both research and legal questions [34].

## 4   Metadata Management

Taking into account the target platform, not all constructs available in our stereo-typed class diagram (such as association classes or composite attributes) may be directly implemented in it [6]. The Model Driven Architect (MDA) approach [26] addresses this problem by defining Platform Independent Models (PIMs) and Platform Specific Models (PSMs), which are automatically obtained from PIMs. Then, the PSMs will be implemented in the target platform.

In order to create the trace database from the stereotyped class diagram (hereafter PIM class diagram), we carry out two schema transformations: (1) the automatic transformation of this PIM class diagram into another stereotyped class diagram closer to the target platform (hereafter PSM class diagram) and (2) the transformation of this PSM to the SQL source code which will generate the database. Next, we explain in detail these two schema mappings.

Firstly, we carry out an exogen transformation [19] since the PIM and the PSM models conform to two different profiled metamodels. For each PIM class diagram created from a statechart, we have identified several types of elements which have to be replaced with other more suitable UML elements in order to obtain the PSM class diagram closer to the target platform. These elements are:

- *Association classes.* Each UML association class is replaced by a new class and two associations. This transformation has obliged us to redefine the SEP profile applying the stereotype «TransitionBetweenStates» to UML classes instead of UML association classes.

- *Composite attributes.* Following our transformation patterns, each class which is the translation of a composite state has a composite attribute for each region. Complex attributes are usually modeled as associations [35], so we propose to replace each of these composite attributes by a bi–directional association between the class which has the composite attribute and the abstract region class which corresponds to the type of the attribute.
- *Inheritance hierarchies.* For the mapping of the inheritance, we have directly focused on the mapping strategies of inheritance hierarchies to databases. There are three fundamental mapping strategies [3,35]: (1) filtered mapping (creating one table for an entire class hierarchy), (2) horizontal mapping (creating one table per concrete class), and (3) vertical mapping (defining one table per concrete class). It is important to note that none of these solutions are ideal for all situations, so the developer has to decide in each case which solution is the most appropriate [3,35]. Considering our particular source class diagram, we have chosen the vertical mapping, given that this option is the simplest, easiest to maintain and least error prone [35].

The second schema mapping deals with the transformation from the PSM class diagram to the SQL source code which will implement the trace database. In order to perform this schema mapping, we have defined several rules concerning the creation of tables, constraints and triggers of the database. In particular, we have defined rules to create two types of triggers, devoted (1) to enforcing in the database specific constraints represented in the PSM class diagram (such as multiplicities) and (2) to the implementation of the specific semantics of UML statechart elements based on the stereotypes applied to the PSM class diagram.

Regarding the methodology followed in this second schema mapping, we want to note that different ways of generating database structures out of class diagrams have been proposed. Nevertheless, the particularity of our approach is that the used class diagrams are stereotyped and each stereotype is translated in a different way, leading to specific triggers for each stereotype.

# 5   Implementation

In order to automatically develop the trace database from a statechart representing a guideline, we have used two different MDA–based tools with support for customizable model–to–model and model–to–text transformations respectively.

On the one hand, to perform the specific model-to-model transformations of our approach (that is, from the UML statechart to the PIM class diagram and from this stereotyped class diagram to the PSM class diagram) we have used an Eclipse plug–in called ATL [5]. ATL is a hybrid model transformation language (it provides both declarative and imperative constructs) and allows developers to design three different kinds of ATL units, of which we use *ATL modules* and *ATL libraries*. ATL defines two types of constructs, rules and helpers, and depending on the chosen programming mode, it distinguishes two different kinds of rules: the *matched rules* (declarative) and the *called rules* (imperative).

On the other hand, to carry out the model-to-text transformation of our approach (that is, from the PSM class diagram to the SQL source code) we have chosen the MofScript tool. MOFScript is, like ATL, another Eclipse plug–in which implements the MOFScript language [20]. Each MOFScript transformation consists of transformation rules which are basically the same as functions, and which define the behavior of the transformation.

By using these two MDA–based tools, we are able to automatically generate the trace database from a statechart.[1] Next, we briefly show how we have used these tools to carry out this task.

**Generating the PIM Class Diagram from a Statechart.** In order to carry out this transformation, our transformation patterns have been implemented into the ATL language. The result of such implementation is the definition of an ATL module (composed of a set of seventeen helpers, five matched rules and twelve called rules) together with two ATL libraries. Then, the defined ATL units take the statechart and the SEP profile as source models. Both models conform to the UML 2.0 metamodel and have been created using the UML 2.0 Eclipse plug–in [10] as .uml2 extension files. By using the defined ATL module together with the two libraries, the statechart is translated into the PIM class diagram which also conforms to the UML 2.0 metamodel.

**Exogen Transformation from the PIM to the PSM Class Diagram.** To carry out this transformation, we have defined another ATL module. This ATL module replaces association classes, composite attributes and inheritance hierarchies from the PIM class diagram with other more suitable UML elements (as we have described in section 4). On the one hand, we consider that the definition of the ATL rules related with the removal of composite attributes and inheritance hierarchy is straightforward and does not require further discussion. In the case of the removal of association classes, we have taken, as a starting point, the predefined `RemovingAssociationClass` ATL transformation from the catalogue of ATL model transformations of [5]. This predefined transformation replaces an association class by a class and two non–directional associations. Since following our transformation patterns, associations in the class diagram are defined as bi–directional associations, we have modified the rules in the `RemovingAssociationClass` transformation so that the new defined associations are bi–directional. Then, this ATL module takes as source models the PIM class diagram and the SEP profile (slightly modified, as we have explained previously), and return the PSM class diagram as a .uml2 extension file.

We would like to point out that during the definition of the ATL units, we experienced various inconveniences associated with the ATL tool. Firstly, we have found several bugs that are not yet resolved. Also, we have had some difficulties concerning the application of profiles to class diagrams. This is due to the fact that, as explained in the ATL Language Troubleshooter [5], with the

---

[1] The transformations are available at www.unizar.es/ccia/articulos.htm

version used, Eclipse UML2 Profiles and Stereotypes can not be applied directly using ATL. We have therefore used the recommended native Java methods of the UML2 plug–in. This, together with the fact that ATL documentation on "how to do" is rather deficient, has made this task a difficult endeavor. Nevertheless, we consider ATL to be a good and easy-to-use MDA tool.

**Generating the SQL Source Code of the Trace Database.** As we have remarked previously, in order to perform the final schema mapping, we have used the MOFScript tool. The SQL code generator is defined as a set of transformations in the MOFScript language. In particular, we have defined three MOFScript files concerning the creation of tables, constraints and triggers respectively. As for the development of the rules concerning the definition of the tables, we have followed the approach of [25] for "UML to RDBMS mapping". For almost each rule proposed in [25] we have defined another rule with the same purpose and functionality. In addition, we have defined other rules to carry out the complete transformation considering specific multiplicities of our PSM class diagram. Regarding the creation of triggers, the defined rules translate each applied stereotype of the SEP profile into the corresponding trigger by following the specific semantics of the corresponding UML statechart element.

The defined MOFScript transformations take the UML 2.0 metamodel as metamodel and the PSM class diagram as the source model and return several `.sql` extension files with the SQL statements which create the tables, constraints, and triggers that constitute the trace database.

We want to note that although there are several tools that generate data models out of class diagrams, the specific issue of our proposal is that each stereotype is translated to a trigger which implements the semantics of the corresponding element of the statechart from which the stereotype comes.

# 6   Related Work

The necessity of storing the trace of systems execution has been stated in the literature within different research fields [12,16,22]. Focusing on the clinical context in general, and on the use of clinical guidelines in particular, we focus on the way in which other decision support systems have tackled the issue of recording the trace of guidelines application. Although during the past thirty years, there have been several efforts to develop tools to realize computer-assisted management of clinical guidelines (e.g. GLIF, Asbru, EON, GEM, PROforma, GLARE) [28], to our knowledge, only GLIF [12] and ASBRU [37] provide mechanisms for recording the execution trace of guidelines to patients. We now compare our proposal with these approaches, basing our comparison on different criteria.

Regarding the strategy chosen to store the application traces, in general, authors use simple textual trace files or databases with only one table [4,14,37]. This fact is clearly a consequence of the consideration of a very simple notion of trace. This is the case of both GLIF and Asbru. On the one hand, Asbru, by means of its Spock model [37], creates for each application of the guideline to a patient, a textual summary by defining *guideline application logs*. On the other hand, in GLIF the execution traces are recorded as XML files conforming to a DTD. When the notion of trace is defined in a more complete way, and consequently in a more complex way, as occurs in our case, the storage structures are not so easy to be determined. One of the advantages of our proposal is that these structures are automatically generated so that the complexity can be managed more easily. Taking into account the recording structures used in other proposals, the benefits of using temporal databases in contrast to other storage structures such as log files [36] are another advantage of our approach.

Another important difference between our approach and the others is related to the method used to create the storage structures. We want to remark that, to our knowledge, neither of the proposals (GLIF and Asbru) uses a MDD approach to create the structure of the execution traces. When this goal is tackled in a MDD setting, the system execution paths can be expressed as behavioral model execution paths [16]. In this respect, the model tracing issue has been analyzed in the literature with respect to different behavioral models: workflows [22], sequence diagrams [16], petri nets [17] or statecharts [16]. In the particular case of statecharts or their UML version (UML state machines), authors define the execution trace in an over simplified way, as a linear sequence of states, transitions and events [7,17]. The problem is that, in contrast to our approach, this definition does not take into account the notions of hierarchy and concurrency, which are two of the main components of the rich expressivity of statecharts. In this way, execution traces are provided with a more complete information.

## 7   Conclusions and Future Work

In this paper we have presented a model-based development approach to developing automatically persistent databases which record the history of the application of clinical guidelines to patients. Starting from the statechart representing a guideline, we propose to carry out several schema mappings to automatically generate the trace database for that clinical guideline. The most significant contributions of this research paper are the following: (1) the way in which the structure of the execution traces is defined, represented not as a linear sequence but as a hierarchy of states; (2) the fact of using a Model Driven Development (MDD) approach in the specific context of healthcare. It must be noted that, to our knowledge, this is the first work to use a MDD based approach for the automatic generation of storage structures for recording the information generated during the application of clinical guidelines. For this reason our approach would open a new field for the application of the MDD approach.

There are several lines of further work. At present, the proposed system does not allow the physician to make decisions which diverge from the strategy defined in the guideline. We think that it would be interesting to extend the system to allow this functionality. Secondly, the study of the way in which the instances generated by the application of the UDSS are recorded in the trace database constitutes an interesting goal for further development. Thirdly, taking into account the possibility of change over time in the definition of a guideline, we consider that the way in which the trace database can be used as a resource for these changes is an interesting issue of future work. Finally, the development of other ways of storage, such as XML documents, deserves to be investigated.

# References

1. Agency for Healthcare Research and Quality. National Guideline Clearinghouse. Guidelines for the prevention of intravascular catheter-related infections, Last visited: January 2008, http://www.guideline.gov
2. Ali, A.B.H., Boufarès, F., Abdellatif, A.: On the Global Coherence of Integrity Constraints in UML Class Diagrams.. In: Databases and Applications, pp. 109–114. IASTED/ACTA Press (2006)
3. Ambler, S.W.: Mapping Objects to Relational Databases: O/R Mapping, (October 2006). Last visited: January 2008,
http://www.agiledata.org/essays/mappingObjects.html
4. Andrews, J.H., Zhang, Y.: Broad-spectrum studies of log file analysis. In: Proceedings of the 22nd international conference on Software engineering (ICSE 2000), Limerick, Ireland, pp. 105–114. ACM Press, New York (2000)
5. ATL (ATLAS Transformation Language) Eclipse Plug-in. Webpage, Last visited: January 2008, http://www.eclipse.org/m2m/atl/
6. Büttner, F., Gogolla, M.: Realizing UML Metamodel Transformations with AGG. Electr. Notes Theor. Comput. Sci. 109, 31–42 (2004)
7. Cavalli, A.R., Gervy, C., Prokopenko, S.: New approaches for passive testing using an Extended Finite State Machine specification. Information & Software Technology 45(12), 837–852 (2003)
8. Dietrich, S.W., Urban, S.D.: An Advanced Course in Database Systems: Beyond Relational Databases. Prentice-Hall, Englewood Cliffs (2005)
9. Domínguez, E., Pérez, B., Rodríguez, A., Zapata, M.A.: Protocolos médicos para la toma de decisiones en un contexto de computación ubicua. Novática 177, 38–41 (2005)
10. EMF-based, U.M.L.: 2.0 Metamodel Implementation. The Eclipse UML2 project website, Last visited: January 2008, http://www.eclipse.org/uml2
11. Tansel, A.U., et al.: Temporal databases: Theory, design, and implementation. Benjamin/Cummings (1993)
12. Wang, D., et al.: Design and implementation of the GLIF3 guideline execution engine. J. of Biomedical Informatics 37(5), 305–318 (2004)
13. Gamma, E., Helm, R., Johnson, R., Vlissides, J.: Design patterns: elements of reusable object-oriented software. Addison-Wesley, Reading (1995)
14. Hu, Z., Shatz, S.M.: Mapping UML Diagrams to a Petri Net Notation to Exploit Tool Support for System Simulation. In: Proc. Int'l Conf. on Software Engineering and Knowledge Engineering (SEKE 2004), pp. 213–219 (2004)
15. Kolaitis, P.G.: Schema mappings, data exchange, and metadata management. In: Proceedings of the 24th Annual ACM symposium on Principles of database systems (PODS 2005), New York, NY, USA, pp. 61–75 (2005)
16. Lencevicius, R., Metz, E., Ran, A.: Tracing Execution of Software for Design Coverage. In: IEEE International Conference on Automated Software Engineering (ASE 2001), pp. 328–332 (2001)
17. Lian, J., Hu, Z., Shatz, S.M.: Simulation-Based Analysis of UML Statechart Diagrams: Methods and Case Studies. The Software Quality Journal (SQJ) (to appear, 2007)

18. McKeeney, L.: Legal issues for the prevention of pressure ulcers. Journal of Community Nursing 16(7), 28–30 (2002)
19. Mens, T., Gorp, P.V.: A taxonomy of model transformation. Electr. Notes Theor. Comput. Sci. 152, 125–142 (2006)
20. MOFScript Eclipse plug in. Last visited: January 2008,
    `http://www.eclipse.org/gmt/mofscript`
21. Murphy, R.: Legal and Practical Impact of Clinical Practice Guidelines on Nursing and Medical Practice. Nurse Practitioner 22(3), 138–148 (1997)
22. Muth, P., Weißenfels, J., Gillmann, M., Weikum, G.: Workflow History Management in Virtual Enterprises Using a Light-Weight Workflow Management System. In: Proceedings of the Ninth International Workshop on Research Issues on Data Engineering (RIDE), pp. 148–155 (1999)
23. Niaz, I.A., Tanaka, J.: Code Generation From UML Statecharts. In: Proc. 7th IASTED Conf. on Software Engineering and Application (SEA 2003), pp. 315–321 (2003)
24. Institute of Medicine. Guidelines for Clinical Practice: from Development to Use. National Academy Press, Washington (1992)
25. OMG. MOF 2.0 Query / Views / Transformations RFP, October 2002. ad/02-04-10, Last visited: January 2008, `http://www.omg.org/`
26. OMG. OMG Model Driven Architecture, June 2003. Document omg/2003-06-01. Last visited: January 2008, `http://www.omg.org/`
27. OMG. UML 2.0 Superstructure Specification, August 2005. Document formal/05-07-04. Last visited: January 2008, `http://www.omg.org/`
28. OpenClinical. Methods and tools for representing computerised clinical guidelines, Last visited: January 2008, `http://www.openclinical.org/gmmsummaries.html`
29. Papadopoulos, C.: The development of Canadian clinical practice guidelines: a literature review and synthesis of findings. J. Can. Chiropr. Assoc. 47(1), 39–57 (2003)
30. Porres, I., Domínguez, E., Pérez, B., Rodríguez, A., Zapata, M.A.: Development of an Ubiquitous Decision Support System for Clinical Guidelines using MDA. In: Proceedings of the CAiSE 2007 Forum, Trondheim, Norway (June 2007)
31. Porres, I., Domínguez, E., Pérez, B., Rodríguez, A., Zapata, M.A.: A Model Driven Approach to Automate the Implementation of Clinical Guidelines in Decision Support Systems. In: Proceedings of the Engineering of Computer-Based Systems (ECBS 2008) (accepted for publication, 2008)
32. Samanta, A., Samanta, J., Gunn, M.: Legal considerations of clinical guidelines: will NICE make a difference? J. R. Soc. Med. 96(3), 133–138 (2003)
33. Selic, B.: The Pragmatics of Model-Driven Development. IEEE Software 20(5), 19–25 (2003)
34. Shahar, Y.: Timing is everything: Temporal reasoning and temporal data maintenance in medicine. In: Horn, W., Shahar, Y., Lindberg, G., Andreassen, S., Wyatt, J.C. (eds.) AIMDM 1999. LNCS (LNAI), vol. 1620, pp. 30–46. Springer, Heidelberg (1999)
35. Sparks, G.: Database Modelling in UML. Spring, issue of Methods & Tools (2001), Last visited: January 2008, `http://oracle.ittoolbox.com/pub/FM041901.pdf`
36. Weikum, G.: Workflow Monitoring: Queries on Logs or Temporal Databases. In: Proceedings of the 6th International Workshop on High Performance Transaction Systems (HPTS 1995), Asilomar (CA) (1995)
37. Young, O., Shahar, Y.: Spock: a hybrid model for runtime application of asbru clinical guidelines. In: Proceedings of MEDINFO-2004, the Eleventh World Congress on Medical Informatics (2004)

# The Research on the Algorithms of Keyword Search in Relational Database

Peng Li, Qing Zhu, and Shan Wang

Key Laboratory of Data Engineering and Knowledge Engineering
School of Information, Renmin University of China, Beijing 100872, P.R. China
{Sapphire,zq,swang}@ruc.edu.cn

**Abstract.** With the development of relational database, people require better database not only in the aspect of database performance, but also in the aspect of the database's interactive ability. So that the database is much more friendly than just before and it is possible for a common user who do not have any special knowledge on database can access the database, without knowing the schema of database and writing intricate SQL. For the reason that the information retrieval on the web has developed well to some extent, when we develop the technology of keyword search in relational database, we can draw some ideas from information retrieve. But the great differences between the text database on the web and the relational database also bring some new challenges: 1) The answer needed by user is not only one tuple in database, but the tuple sets consisting of the tuple connect from different table using the "primary key-foreign key" relationship. 2)The results of the evaluation criteria is more important, because it is directly related to the effectiveness of Search System. 3)The structure of relational database is much more intricate than text database, and the algorithms of information retrieval are not fit the relational database. So in this paper, we introduce a novel keyword search algorithm and a modified criteria of evaluating answers in order to enhance efficiency of the keyword search and return much more effective information to users, finally, the search algorithm's performance is tested and evaluated.

**Keywords:** Keywords search, relational database, information retrieval, search algorithms.

## 1 Introduction

As we know, the demand for user to search in relational database with keywords is growing steadily, The problem we have encountered is we should design a novel algorithm to make the users to search in the relational database with keywords efficiently and we should design a fit Ranking-Mode to return more effective top-k answers to users. Now we introduce the problem with a practical instance.

There are four tables in the DBLP database and the lines with an array shows the relationships. If user want to get the papers published by famous database

Y. Ishikawa et al. (Eds.): APWeb 2008 Workshops, LNCS 4977, pp. 134–143, 2008.

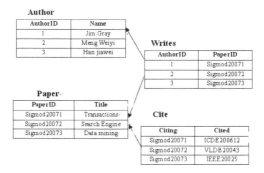

**Fig. 1.** DBLP database

expert "Jim Gray", he/she should learn the schema of the DBLP database firstly, and then he/she must write intricate SQL like this:

*Select Title*
*From Paper, Writes, Author*
*Where Paper.PaperID=Writes.PaperID*
*And Writes.AuthorID=Author.AuthorID*
*And Author.Name=Jim Gray*

It is too hard for a common user, so our aim is that as long as the user submits keywords, the system will return the top-k effective answers to user automatically. Keyword search is very success in web search engine, when user search with the web search engine such as *Google*, *Baidu* and *Yahoo*, What he/she should do is submitting the keywords to the interface provided by search system.

The contributions of our paper are as follows: First of all,we introduce a novel search algorithm, compared to the existing algorithms, the mainly used data structure in our algorithm is sets, and the graph is only the accessorial structure, we only use the depth-first algorithm of Graph to adjust the orders of the elements in the sets to make sure that most adjacent Table in the sets have a"Primary key-Foreign key" relationship, and we can join the tuples of adjacent Tables expediently and construct the final results -Tuple-Tree. So compared to existing algorithms, the time and space cost is much smaller.

Secondly, we have done with the special terms, there are two kinds of special terms in the keywords: Schema-Terms, Redundant-Terms. We add these peculiar term processing to make the results more effective the experiments show that our idea is feasible.

Thirdly , our new search algorithm can effectively limit the size of the results, because when a user input keywords, the system generate a serial of Table name sets, the max size of these sets are limited to the twice number of Tables to avoid additional cost and make our algorithm more effective.

In Section 2, we discuss the related work. In Section 3, we introduce the system model such as Answer-Graph and Tuple-Tree. In Section 4, we describe the design of our algorithms. In Section 5, we present experiments results and analysis them to demonstrate the effectiveness of our results generated by our algorithms. Section 6 draws the conclusion and outlines directions for future work.

## 2   Related Work

The analysis of the state-of-art of the information retrieval (IR) shows that apply information retrieval engine to relational database is a challenging work, because there is a great difference between text database and relational database. What information retrieval search in is text databases, their meta information is documents. As soon as a user input the keywords, the search engine will calculate the similarities between the keywords and documents, and then the search engine orders the documents with their similarity to keywords in descendent, finally the search engine return the top-k answers to user. But in relational database, the data organized in the form of various tables, data is stored in these tables, and the tables are connected with "primary key-foreign key" relationships. Obviously, the information contained in each tuple is limited, we focus on how can we join the tuples into a Tuple-tree that contains the information which user want to know.

The keywords search systems in relational database so far have primary keyword search functions. For the system named BANKS[1]: It uses directed data graph to help with its keywords search, in its data graph, if two tuple's node have a "primary key-foreign key" relationship, there are two corresponding directed edges between them, and it utilize the reverse heuristic search algorithm to find the node containing information; DBXPlorer[2] uses undirected graph to help with keyword search, first of all, it accesses to symbol table to get tuples' information , and then calculates Tuple-Tree according to Schema-Graph, at last it constructs each SQL statements according each Tuple-Tree; while DIS-COVER[3] executes the searching function using a complicated data structure named Tuple-Graph, and supported "And" and "Or" semantics. In general, their idea is that they construct a Answer-Graph preliminary, when the user submits the keywords, they choose the minimum sub-graph that contains all the terms in keywords. They did not draw any idea from information retrieval[11], the way of generating answer is too simple and has a high cost, and the Ranking-Model dos not fit users' wills to some extent, especially, they often return the most complicated answer to user directly.

## 3   System Model

### 3.1   System Framework

First of all , we get the keywords submit by users and filter out the redundant words such as "and" , "the" , "by", then the query mode access to the database with these keywords and get two kinds of information: 1. Schema-Graph of the database; 2. Table-Tuple list acquired with keywords and the corresponding tuple sets containing one or more terms of the keywords. Now according to Table name and Table-Tuple list we can get corresponding tuple sets, join these tuples iteratively with "primary key-foreign key" relationships to get a lot of tuple trees. At last , we rank these Tuple-Trees according to the similarities between Tuple-Trees and keywords, return to user the top-k answers.

## 3.2 Concepts and Terminology

**Schema-Graph.** Relational databases are based on relational mode, it describes world with relationships. A relationship not only can be used to describe entity and its attributes, but also can be used to describe relationships between entities. In fact, Schema-Graph is a model of directed graph, each node represents the name of a table, if two table has a "primary key-foreign key" relationship, there is a corresponding line with an array between two nodes representing these two tables. The Figure 1 is an instance of Schema-Graph[4].

**Answer-Graph.** A Answer-Graph is the sets consisting of Table names, given the keywords, system will scan the database, if tuple in one Table contains a term of the keywords, add the Table name into Answer-Graph. After finishing generating Answer-Graph, we change the order of the elements in Answer-Graph by a depth-first traversal of Schema-Graph to make sure that every two Table in the Answer-Graph has a "primary key-foreign key" relationship.

**Table-Tuple list.** Table-Tuple list is an array of (Table name, Tuple) pairs. In the progress of scanning basic Table, if the tuple contains one term of the keywords, record the Table name and this tuple, they form a (Table name, Tuple) pair, then update the Table-Tuple list, the update way is divided into two situations: 1. If the Table-Tuple list has not been constructed, make the (Table name, Tuple) pair to be the first element of the Table-Tuple list. 2. If the Table name in (Table name, Tuple) pair has already existed in Table-Tuple list, add (Table name, Tuple) pair to the list. In the progress of scanning foreign Table, every tuple in foreign table will form (Table name, Tuple) pair, and all of them will be added to the Table-Tuple list.

**Tuple-Tree.** Tuple-Tree is generated according to Answer-Graph, its abstract logical structure is similar to Answer-Graph, but its elements are not Table names but are some tuples coming from different Tables[9]. It detailed definitions are as follows:

1) If a node in Tuple-Tree are one of tuples in basic Table, it contains one or more terms of the keywords.

2) There is no reduplicate tuple in the Tuple-Tree.

**Schema-Terms.** Terms that same to the names of database, or the names of Table and attributes, we call these terms "Schema-Term". In the database showed in Figure 1, if terms are "Author", "Paper", "Name" and so on, they are Schema-Terms, they are rare in Tables, so when we consider them, their weights are thought to be very low even equal to zero. But it is not appropriate for the reason that these words implicitly shows that the query is inclined to get some Tables' information or some attributes.

**Redundant-Terms.** In this paper, we have also taken a special case into account: There are a sort of special terms in keywords and Tuple-Trees, such "The", "by", "and", they are indispensable, but they do not have any real meaning but their occurrence frequencies is very high.

**Instance.** In this part, we use an instance to illustrate these system models, when a user submit *"Sigmod MengWeiYi"* as keywords to system, at the very start, system analysis keywords to realize that there are two terms in keywords. And then the system will scan the database , Table *Author* is in possession of the tuples contains the terms *"MengWeiYi"* Table *Paper* owns the tuples contains "Sigmod". As a result, the system will generate two sets: *Paper* and *Author* as the Answer-Graphs. Both of them are the names of basic Tables who are in possession of one or more terms of keywords. The objects which should be analysised recursively by the system are these sets contains Table names, the analysis process will be ended until the number of sets to be analysised is limited to one. So the system will analysis sets *{Paper}* and *{Author}*. For the reason that Table *Paper* and Table *Author* have common foreign Table *Write*, the system generate a new set *{Paper, Paper-Author, Table}* as another Answer-Graph. Now the system will be ready to generate Answer-Graph that the elements may be reduplicate for the reason that the Tuple-Trees' information should originate from same table sometimes. The system combines the existing sets *{Paper}* and *{Paper, Write, Author}* to get a new set *{Paper, Write, Author, Paper}*. Because the Table *Paper* and Table *Author* have common foreign Table *Write*, the system add element "Write" into this set and reorder the elements of this set in order to make sure that every adjacent elements own a "Primary key-Foreign key" relationship. As a result the newly generated set is *{Paper, Write, Author, Write, Paper}*. In this way we can also get another new set *{Author, Write, Paper, Write, Author}* by combining set *{Author}* and *{Paper, Write, Author}*. These two sets are owns the same elements so they are in the same on earth. So our system will make a random choice as a new set from these same sets. Finally our system will generate four Answer-Graphs. These Four Answer-Graphs are shown in Figure 2.

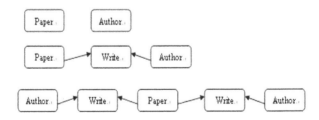

**Fig. 2.** Four Answer-Graphs

We can get the Answer-Graphs and corresponding Table-Tuple lists at the same time, the three corresponding Table-Tuple lists are illustrated in Figure 4.

In scanning Tuple-Name lists, we can get corresponding tuples, the next work is join them according "Primary key-Foreign key" relationship, the Tuple-Trees are as follows:

Tree1: *32 MengWeiYi*
Tree2: *Sigmod20071 DataStreams*
Tree3: *32 MengWeiYi 32Sigmod20071 Sigmod20071 DataStreams*
Tree4: *32 MengWeiYi 32 Sigmod20071 Sigmod20071 DataStreams 59 Sigmod20071 59 ElementYu*

We will return them to users after ranking them.

## 4    Algorithms and Ranking-Model

Our keyword search system in relational database owns two core algorithms: The first one is the algorithm of generating Answer-Graph; The second one is Analysising the Answer-Graph and then generating Tuple-Tree. Both of them are in the Query-processing module. We will discuss two algorithms in detail.

### 4.1    Generate Answer-Graph

The idea of this algorithm is scanning the database at first to get Table-Tuple list and initial Answer-Graphs. These initial Answer-Graphs are a serial of Table name sets, each sets contain only one basic Table name. And then analysis these basic Table names, check every pair of them, if they have common foreign Table according to Schema-Graph, create a new set, add them and their common foreign Table name to this set. These newly constructed sets are the object we analysis in the next recursive time and we update Answer-Graph at once. In the next recursive time, we check these sets again, if the elements of a pair of sets have common foreign Table, combine these two sets into one to form a new set again. After generating the sets owns no reduplicate elements, we generated new sets contains reduplicate elements by combining existing sets and the size of the new set are limited to the twice of the number of the Tables in database. This function will recursively use this algorithm till the number of newly constructed sets which will be analysised is limited to one. The algorithm is illustrated in Figure 3.

### 4.2    Generate Tuple-Tree

The input of this algorithm is the initial Answer-Graph, the elements of them is out of order. That is to say, every two Table in the initial Answer-Graph may have no relationship according to Schema-Graph, So if the size of Answer-Graph is larger than 2, we will convert the order of the elements in Answer-Graph with depth-first traversal of Schema-Graph so as to make the every adjacent elements have a "Primary key-Foreign key" relationship. In this way, it is much easier to generate Tuple-Tree by joining tuples together and the times of Scanning table and Table-Tuple list can be greatly reduced.The algorithm is illustrated in Figure 4.

```
void Algorithms::GenAG(Initial Answer-Graph AG)
1. Scan the database, Create Schema-Graph
and get Table-Tuple list
2. Analysis the Table names in AG
Answer-Graph AGTemp;
if(AG.size==1 or 0)
    return;
else
    for(i=0 to AG.size)
        for(j=0 to AG.size)
        if(AG[i] and AG[j] are adjacent)
            Combine AG[i] and AG[j]
            and their Common Foreign-Table into a new set temp;
            Add temp to current Answer-Graph AGTemp;
            Create Tuple-Tree;
        Analysis(AGtemp)
    return;
```

```
1. Convert Answer-Graph to make foreigh-table to be the first
element so that the deep-first traversal more effectively
2. Deep-first traverse the Answer-Graph and reconstruct the
original Answer-Graph
3. Scan the Tuple-Tree, inqury the Name-Tuple list and get the
Tuples corresponding to the first element of Answer-Graph, and
the tuple's name is added to temp1.
4. Scan Temp1;
    Arrange the tuples adjacent to temp1 to temp2;
    Scan temp2;
        if(temp1[m] and temp2[n] can be joined)
            join this two tuples
    update temp1 and temp2;
    Continue the join action util finish scanning
    End Scan
5. Return
```

**Fig. 3.** Generate Answer-Graph          **Fig. 4.** Generate Tuple-Tree

### 4.3   Ranking-Model

**Our Ranking-Model.** As the Ranking-Model in information retrieval is relatively mature, we try to draw the ideas from Ranking-Model in information retrieval.

We use classic vector space mode in IR[8] and compare each Tuple-Tree returned by our algorithms with the document returned as answer in information retrieval, so the formulas which compute similarity between keywords and Tuple-Tree are as follows[5]:

$$Weight(Q, T) = \sum Weight(k, Q) * Weight(k, T) \tag{1}$$

$$Weight(k, T) = \sum \frac{Weight(T_i)}{Nsize(T)} \tag{2}$$

$$Nsize(T) = 1 - s + s * \frac{size(T)}{avgsize} \tag{3}$$

$$ntf = 1 + \ln(1 + (tf)) \tag{4}$$

$$itrf = \ln(\frac{N}{1 + trf}) \tag{5}$$

$$ntrf = (1 - s) + s * (\frac{s * trl}{avgtrl}) \tag{6}$$

*Term frequency(tf)*: refers to the frequency that a term appears in the Tuple-Tree. As we know, the more times a term appears in the Tuple-Tree, the larger weight it should own. But if a Tuple-Tree has a large scale of tuples, some term may appear in it so many times, the weight of this term may at a linear increase, the system will be bias at only returning large-scale Tuple-Tree only, that is not appreciate enough to return effective results. So we should have the *Term frequency(tf)* normalized.

*Tree frequency(trf):* refers to the number that the Tuple-Tree which contains a term of keywords in a collection. The greater of the *Tree frequency(trf)*, the lower weight of the corresponding term. So when we compute the weight of a term, we consider the *Tree frequency(trf)* of the term as the divisor.

*Tree length(trl):* refers to the number of terms that in a Tuple-Tree. We confront the same problem when we calculate it as we calculate the *Term frequency(tf)* that a large-scale Tuple-Tree may has too many terms, and the *Tree length(trl)* calculated according to this type of tree is not appreciate either. We should normalize it according the formula (6). The *avgtrl* is the average value of the *Tree length(trl)s* of all the terms in keywords. The value of s is 0.2.

**Ranking-Criteria.** We should choose a scientific and rational Ranking-Criteria to test the effectiveness of the results returned to users and we use *recall* and *precision* and the combination of them-*RP-measure*, it is a classical way in information retrieval the formulas are as follows[10]:

$$P = Precision = (\frac{relevant retrieved}{retrieved}) \tag{7}$$

$$R = Recall = (\frac{relevant retrieved}{all}) \tag{8}$$

$$result = \frac{2 * P * R}{P + R} \tag{9}$$

We set a threshold $A=1.5$, only if the result of a Tuple-Tree calculated in formula (11) is larger than $A$, we consider this Tuple-Tree as an relevant result.

## 5   Experiments

### 5.1   Experiments Configuration

The experiment is on the study of algorithm simulation, so we choose file system to simulate database, the data source we choose is DBLP database owns 400000 tuples, we download it from *http://dblp.uni-trier.de* and the data format of this database is XML, we convert the data format into a novel format that our system can copy with. Our experiment is implemented in C++ and run on a PC with AMD athlonX2 CPU and 2GB ram.

### 5.2   Experiment Results and Analysis

First of all the system generate 50 lines of keywords in a random, and then query the DBLP database with these keywords. There are 5 DBLP databases to be test by our system and their size are ranging from 200K to 1500K.The default semantic supported by system is *Or*, the keywords are considered to be term1 OR term2 OR term3......

**Experiment 1:** This experiment test the influence that the system does with Schema-Terms additionally, we also calculate the average effectiveness of the results returned by 10 lines of keywords tested on 5 DBLP databases, the results are illustrated in Figure 5.

After adding Schema-Term processing function to system, the weight of Tuple-Tree that contains Schema-Graph will be higher than before, and they will be more likely returned to users.

**Experiment 2:** This experiment test the influence that the system do with Redundant-Terms additionally, we also calculate the average value of results' effectiveness returned by 10 lines of keywords tested on 5 DBLP databases, the results are illustrated in Figure 6.

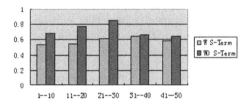

**Fig. 5.** Results' effectiveness based on different Schema-Term processing

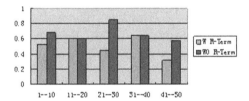

**Fig. 6.** Results' effectiveness based on different Redundant-Term processing

These results show that when there is no Redundant-Terms in query, the Redundant-Term processing takes no effect. But if there is any Redundant-Terms in the keywords, the improvement is very apparent, because the Tuple-Tree with large scale may have too many Redundant-Terms.

In the experiment, we have encountered two problems: 1. The Ranking-Model is not mature enough to some extent, sometimes the returned results are still useless, and the effectiveness is at a low level. We think that we should design novel Ranking-Model. 2. There are many phrases in the keywords that users submit, we will focus on how to identify these phrases in keywords and Tuple-Trees, and add phrase processing module to the system in order to improve the effectiveness of the results.

## 6   Conclusions

It is a attempt for us to propose a novel search algorithm not only take the advantage of sets' convenience but also own the logical structure of directed graph in that the Schema-Graph can help with adjusting the order of the elements in sets. So there is a good balance between effectiveness and efficiency of our algorithm. After adding special case processing module such as Schema-Term processing and Redundant-Term processing, when query in the database with a large scale, the answer will be much more effective than before. But our algorithm is still immature, we will focus on the study of designing novel Ranking-Model, adding phrase processing algorithm and Pseudo Relevance Feedback module to filter the answer again to return best top-k answers to users.

## Acknowledgment

This work is supported by National Natural Science Foundation of China under Grant No. 60473069, Science Foundation of Renmin University of China (No. 30206102.202.307), and the project of China Grid (No.CNGI-04-15-7A).

## References

1. Agrawal, S., Chaudhuri, S., Das, G.: DBXplorer: A System For Keyword-Based Search Over Relational Databases (2002)
2. Hulgeri, A., Bhalotia, G., Nakhe, C., Chakrabarti, S., Sudarshan, S.: Keyword Search in Databases (2001)
3. Hristidis, V., Papakonstantinou, Y.: DISCOVER: Keyword search in relational databases (2002)
4. Hristidis, V., Gravano, L., Papakonstantinou, Y.: Efficient IR-Style Keyword Search over Relational Databases. In: VLDB 2003 (2003)
5. Singhal, A., Buckley, C., Mitra, M.: Pivoted Document Length Normalization. In: SIGIR 1996 (1996)
6. Liu, F., Yu, C., Meng, W.: Effective keyword search in relational database. In: SIGMOD 2006 (2006)
7. Liu, S., Liu, F., Yu, C.T., Meng, W.: An effective approach to document retrieval via utilizing WordNet and recognizing phrases. In: SIGIR 2004 (2004)
8. Salton, G., McGill, M.: Introduction to Modern Information Retrieval. McGraw-Hill, New York (1983)
9. Bast, H., Majumdar, D., Schenkel, R., Theobald, M., Weikum, G.: Io-top-k: Index-access optimized top-k query
10. Chakrabarti, B.K., Ganti, V., Han, J., Xin, D.: Ranking objects by exploiting relationships: computing top-k over aggregation. In: SIGMOD Conference, pp. 371–382 (2006)
11. Singhal, A.: Modern information retrieval: A brief overview of Data processing. In: VLDB 2006, pp. 475–486 (2006)

# An Approach to Monitor Scenario-Based Temporal Properties in Web Service Compositions*

Pengcheng Zhang[1], Bixin Li[1,**], Henry Muccini[2], and Mingjie Sun[1]

[1] School of Computer Science and Engineering
Southeast University, Nanjing, China
[2] Dipartimento di Informatica, University of L'Aquila
Via Vetoio, L'Aquila, Italy
{pchzhang,bx.li}@seu.edu.cn,
muccini@di.univaq.it

**Abstract.** Keeping composite services satisfying desired properties has been widely accepted as an important and challenging problem due to the dynamically evolving attribute of web service compositions. Runtime monitoring and dynamic verification techniques become first class activities to be performed during the execution of web service compositions. Scenario-based temporal properties depicting the complex interactions among the different services are a kind of important property which needs to be monitored at runtime. However, some complex scenario-based temporal properties cannot be easily represented by the traditional formalism such as temporal logic. In this paper, we first propose to represent the scenario-based temporal properties of web service compositions by the use of a novel notation (Property Sequence Chart). Then, we use Aspect-Oriented Programming techniques to extend the open-source BPEL engine (ActiveBPEL) and monitor its execution. Based on these assumptions, we propose a more intuitive approach to monitor the scenario-based temporal properties in web service compositions.

**Keywords:** Web service, Web service compositions, Runtime monitoring, Property Sequence Chart.

## 1 Introduction

One key attribute of web service is that it can support rapid, low-cost and easy composition of distributed applications even in heterogeneous environments [1]. However, keeping composite services satisfying desired properties has been widely accepted as an important and challenging problem due to the dynamically evolving

---

* The work is supported partially by the National Natural Science Foundation of China under Grant No. 60773105, partially by the Natural Science Foundation of Jiangsu Province of China under Grant No.BK2007513, partially by National High Technology Research and Development Program under Grant No. 2007AA01Z141 and partially by the Program for New Century Excellent Talents in University under Grant No.NCET-06-0466.
** Corresponding author.

Y. Ishikawa et al. (Eds.): APWeb 2008 Workshops, LNCS 4977, pp. 144–154, 2008.
© Springer-Verlag Berlin Heidelberg 2008

attribute of web service compositions. One main cause of the dynamic attribute is that composite services are finished at runtime. Even if they are verified for correctness before runtime, composite services may not hold at runtime. For example, they are autonomously developed by third-party developers, and can be changed for reasons such as version update without notifying service requestors.

Therefore, using traditional static validation and verification techniques, such as testing and model checking, can be not enough to keep the correctness of web service compositions. Runtime monitoring and dynamic verification techniques become first class activities to be performed during the execution of web service compositions. Runtime monitoring consists of collecting data about a monitored service and checking whether these data conform to expected properties; it can complement traditional static verification techniques to improve the quality of web service.

In this paper, we consider web service compositions as BPEL [3] specification and use Aspect-Oriented Programming [4] techniques to extend the open-source BPEL engine (ActiveBPEL [5]), and then monitor its execution. The properties of web service compositions can be of various nature (e.g., security, reliability, performance, usability and others). When many different services are composed, the scenarios may be very complex and there are some potential errors among interactions. It will be worse when concurrent clients invoke the same service. Therefore, this paper monitors the scenario-based temporal properties identifying how BPEL processes and their partners are supposed to interact in order to satisfy predefined properties.

The rest of the paper is organized as follows: Section 2 presents the VOS example to motivate the paper. Section 3 shows how to use a novel scenario-based language Property Sequence Chart (PSC) [6, 7] to represent properties. Section 4 describes the framework for runtime monitoring of scenario-based temporal properties of web service compositions. Section 5 provides a description of related work and concludes the paper.

## 2   A Motivation Example

The Virtual Online Shop (VOS) is a composite service that composes a Bank service and a Shop service to provide a combination of sell and payment service [2]. The VOS becomes active upon a request for an item ([RECEIVE]itemRequest). Then, the VOS interacts with a Store to find the corresponding item ([INVOKE]findItem). If the requested item is not available, the Client is informed about ([INVOKE]notAvailable) and the process terminates. Otherwise, the VOS sends an offer with a cost to the the Client ([INVOKE]offerCost), then, it stops and waits for either a positive reply from Client ([ON MESSAGE]getOrdererData), or a negative response ([ON MESSAGE]-offerNack). The Client can also ask for another offer ([ON MESSAGE]offerChange). If he accepts the offer, VOS will pick [ON MESSAGE]getOrdererData from the Client and will start to interact with the Bank. VOS first sends [INVOKE]startPayment to the Bank, and then the Bank will reply with positive acknowledgement or negative one, i.e. [ON MESSAGE]startPaymentAck or [ON MESSAGE]startPaymentNack. If the response is positive, it means that the bank account information for the Client is correct and the VOS can do the payment ([INVOKE]cofirmPayment). If the money in the account is not enough, the VOS can also cancel the payment via [INVOKE] cancelPayment. If the payment is successful ([ON MESSAGE]confirmPaymentAck), the Client is notified with an acknowledgement ([INVOKE]getOrdererDataAck), otherwise, if the payment fails ([ON MES-SAGE]confirmPaymentNack), a refusal message ([INVOKE]getOrdererDataNack) is sent to Client.

After critically analyzing the BPEL process, we can identify four different scenario-based temporal properties according to the BPEL process described above.

**Prop1:** When receiving the item request from a Client, the VOS must immediately interact with the Store service to find whether or not the item is available. Then, if the VOS makes an offers cost to the Client, he must send getOrdererData to the VOS on the condition that he neither sends message offerNack nor sends message offerChange to the VOS.

**Prop2:** When sending getOrdererData message to the VOS, the Client will receive getOrdererDataAck unless the following events are exchanged first: the VOS sends startPayment to the Bank, and the Bank replies with statPaymentAck. Then, the VOS begins to send confirmPayment to the Bank, and the Bank replies with confirmPaymentAck.

**Prop3:** After receiving the getOrdererData message, the VOS invokes the Bank service with message startPayment, and the Bank responses with startPaymentNack. In this case, if the VOS still invokes cofirmPayment, an exception will happen.

**Prop4:** Two Clients can interact with the VOS concurrently. If itemRequest(item1) happens before itemRequest(item2), the response to itemRequest(item1) must happen before the response to itemRequest(item2).

**Prop1**, **Prop2** and **Prop3** deal with one Client interacting with the VOS. **Prop4** is an example of properties on two Clients interacting with the VOS concurrently. Traditional monitoring approaches based on temporal logic need to use temporal logic to represent these properties and translate them into automata. The main monitoring idea is to judge whether the current execution is the input language of the automata. However, it is a difficult task to accurately and correctly express these temporal properties in temporal logic. For instance, a Linear Temporal Logic (LTL) formula representing **Prop4** is shown as follows:

[ ] ((([RECEIVE]itemRequest(item1) & !([INVOKE]offCost(item1)⊕[INVOKE]notAvailable(item1)) U ([RECEIVE]itemRequest(item2) -> <> (([INVOKE]offCost(item1)⊕[INVOKE]notAvailable(item1)) & <> ([INVOKE]offCost(item2)⊕ [INVOKE]notAvailable(item2))))).

Where [ ], <> and U are the temporal operators which mean Globally, Eventually and Until respectively, and a⊕b = (a & !b) || (!a & b) means that either a happens or b happens.

The LTL formula is not easily understandable and even more difficult to write correctly without some special experiences. In the next section, we will propose a more intuitive and closer to natural language description approach to represent these properties, and propose a more intuitive way to monitor these properties.

## 3 Using PSC to Represent the Scenario-Based Temporal Properties of Web Service Compositions

PSC is an extended graphical notation of a subset of UML 2.0 Sequence Diagrams, which is originally used for specifying the interaction between collections of component instances that execute concurrently [6,7]. There are three types of messages: *Regular*, *Required* and *Fail*. Regular messages (labeled with *e:msg* name) constitute the precondition for a desired (or an undesired) interaction. Required messages (labeled with *r:msg* name) must be exchanged by the system and are used to express mandatory interactions. Fail messages (labeled with *f:msg* name) should never be exchanged and are used to express undesired interactions. A *strict* operator is used to explicitly specify a strict ordering between a pair of messages. A *loose* ordering, instead, assumes any other messages can occur between the selected ones. *Constraints* are introduced to define the restrictions on what can happen between the messages containing the constraints and its predecessor or successor. Constraints are classified into three kinds: past constraints, present constraints, and future constraints. Past constraints and future constraints can be both chain and Boolean formulae. In contrast, present constraints can be only Boolean formulae. Boolean formulae constraints are a Boolean expression of messages while chain constraints are a sequence of messages. PSC also uses the *parallel*, *loop*, *simultaneous*, and *complement* operators for specifying parallel merge (i.e., interleaving), iteration, simultaneity and complementary (i.e., all possible messages but a specified one), respectively. The expressiveness of PSC is measured by property specification patterns.

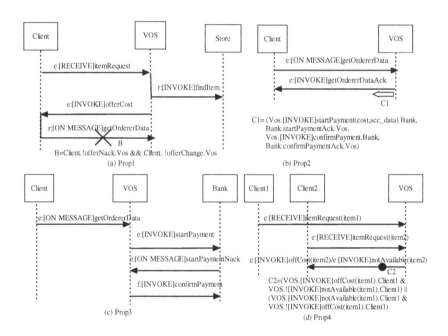

**Fig. 1.** The PSC representations of four different properties

PSC can be used to model the interactions among BPEL process and its partners. BPEL process is modeled by an instance, and all the partners involved in the interactions in the BPEL process are also modeled as other instances. The interaction properties between BPEL process and its partners are represented by PSC. The four properties in Section 2, represented by PSC, are shown in Figure 1. (a), (b), (c) and (d) respectively.

There is a *present constraint* of [ON MESSAGE]getOrdererData in **Prop1**, and a *past chain constraint* of message [INVOKE]getOrdererDataAck in **Prop2**. There is an error message [INVOKE]confirmPayment in Prop3. There is a *past Boolean constraint* in **Prop4**, which means $C2$ must hold before message [INVOKE]off-Cost(item2) or [IN-VOKE]notAvailable(item2) happens. While the LTL formulas are not easily understandable, the same property expressed in the PSC formalism (Figure 1) appears more intuitive and closer to the natural language description.

# 4   The Runtime Monitoring Framework

In this section, we propose a framework to monitor these properties in web service compositions. A similar method is used in [14] to validate the properties for dynamic component-based system. The architecture of the framework is based on the dynamic aspectization of the open-source BPEL engine. ActiveBPEL is instrumented using the Aspect Oriented Programming language (AOP) as AspectJ[8]. In this way, we can add monitoring facilities into an existing application without affecting its execution logic. We can keep business logic and monitoring logic separate, and good code modularization. Our runtime monitoring framework is composed of **Extended BPEL engine with Interceptor**, **Property Database**, **Observer, and Analyzer**.

Our approach uses AOP technique to extend **AcitveBPEL engine with Interceptor**. Obviously, Interceptor which is the core component in our framework allows for defining monitor-related pointcuts and advices in AspectJ. Our approach can define pointcuts for the following events of the engine: Engine start or stop, BPEL process construction or destruction, and key activities of BPEL process such as invoke, receive, and reply. For example, in order to monitor the whole activities exchanged by the composite service VOS, we can define the following aspect in AspectJ to record the activities in a log file.

```
public aspect AOP_Log {pointcut log_virtualonlineshopInterface():execution(*
        VirtualOnlineShop..*(..));
    before ():log_virtualonlineshopInterface () { Signature s=thisJoinPoint.getSignature();
            LogFile.println("[Monitor LOG] Entering: "+s.toString());  }
    after ():log_virtualonlineshopInterface () { Signature s=thisJoinPointStaticPart.getSignature();
            LogFile.println("[Monitor  LOG] Exiting: "+s.toString()); } }
```

The above code shows how to record each activity before it begins or after it has been finished (through before () and after ()). Some activities, such as invoke, have input and output parameters, so we need to monitor it before it begins and after it has been finished.

The first message of each predefined property represented by PSC is identified. Then, four different functions of the first message are defined, i.e. *Constraint, NextCorrect, NextFail*, and *NextIgnore*. These functions will be used by the Analyzer. The messages constrained by the present constraints are contained in a *Constraint* function; the next expected messages are contained in *NextCorrect* according to the PSC; the next unexpected messages are contained in *NextFail*; the next ignored messages are contained in *NextIgnore*. The four functions of each occurred message in PSC can be counted iteratively, and are placed into a **Property Database**.

Our general assumption is that $a$ is the current message in PSC and $b$ is the next message in PSC, and $C$ is the *Constraint* of $a$. If $C$ is not empty, we need to record it in a message set, which will be used by the Analyzer in the future. The other three different functions can be explained according to the following four different conditions. Condition 1 means that the timeline is loose and the type of message is regular or required. Since *b* is the next correct message in PSC, the *NextCorrect(a)* is *{b}*. Because the timeline is loose and permits any other messages exchanged before it, the *NextFail(a)* is empty, and the *NextIgnore(a)* is the complement of *b*, i.e.*{!b}*. Condition 2 means that timeline is loose and the type of message is fail. Since *b* has a loose timeline, the *NextFail(a)* is *{b}* and the *NextIgnore(a)* is *{!b}*. Condition 3 means that the timeline is strict and the type of message is regular or required. If correct message *b* does not happen, the system will raise an error, i.e. the *NextFail(a)* is *{!b}*. Condition 4 means the timeline is strict and the type of message is fail. If *b* is exchanged in the next time, the system will raise an error and all other messages are correct. Note that there are some differences between Condition2 and Condition4. If other messages are exchanged between *a* and *b* on Condition2, the system will also raise an error after *b* happens in the future because the timeline is loose. In other words, it means the error can be delayed. Therefore, the *NextCorrect(a)* is empty and the *NextIgnore(a)* is *{!b}*. If other messages are exchanged between *a* and *b* on Condition4, the system will not raise an error after *b* happens for the strict timeline. It means the error cannot be delayed. So the *NextCorrect(a)* is *{!b}* and the *NextIgnore(a)* is empty.

Condition1 :

$type(timeline) = loose$ & $type(message) = regular \parallel required$

$\begin{cases} NextCorrect(a) = \{b\} \\ NextFail(a) = \varnothing \\ NextIgnore(a) = \{!b\} \end{cases}$

Condition2 :

$type(timeline) = loose$ & $type(message) = fail$

$\begin{cases} NextCorrect(a) = \varnothing \\ NextFail(a) = \{b\} \\ NextIgnore(a) = \{!b\} \end{cases}$

Condition3 :

$type(timeline) = strict$ & $type(message) = regular \parallel required$

$\begin{cases} NextCorrect(a) = \{b\} \\ NextFail(a) = \{!b\} \\ NextIgnore(a) = \varnothing \end{cases}$

Condition4 :

$type(timeline) = strict$ & $type(message) = fail$

$\begin{cases} NextCorrect(a) = \{!b\} \\ NextFail(a) = \{b\} \\ NextIgnore(a) = \varnothing \end{cases}$

**Table 1.** The four different functions of the corresponding properties

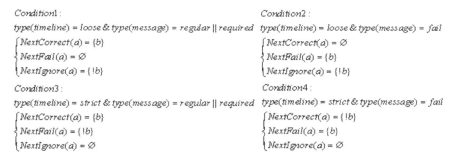

Table 1 shows four different functions for the properties of the VOS example. The first to third messages in **Prop1** have no constraints, so the constraint function is empty. Then, we need to count the other functions. For the first message, the next message of first message ([RECEIVE]itemRequest) has a strict timeline and is required, so we can count it according to Condition3. The second and third message can be done in similar way. Concerning the last message, it has a present Boolean constraint, so we need to record the Boolean expression in function *Constraint* which will be used in the analyzing process. The other three functions of the last message are all empty. Finally, the four functions of messages in **Prop1** are counted. The functions for other properties can be counted in the same way.

The **Observer** component collects the runtime data from the Interceptor. Then, the Observer deals with the collected information and represents them as an easy format which can be recognized by the Analyzer component. Because all the activities of our example just have input parameters, these activities can be intercepted when they begin. One possible execution sequence of the VOS can be observed in Sequence1. Then, if two Clients invoke the VOS service concurrently, a possible execution trace can be monitored in Sequence2.

**Sequence1:**
[Monitor LOG]Entering: [RECEIVE]itemRequest
[Monitor LOG]Entering: [INVOKE]findI tem
[Monitor LOG]Entering: [INVOKE]offer
[Monitor LOG]Entering: [ON MESSAGE]getOrdererDate
[Monitor LOG]Entering: [INVOKE]startPayment
[Monitor LOG]Entering: [ON MESSAGE]startPaymentAck
[Monitor LOG]Entering: [INVOKE]confirmPayment
[Monitor LOG]Entering: [ON MESSAGE]confirmPaymentAck
[Monitor LOG]Entering: [INVOKE]getOrdererDateAck

**Sequence2:**
[Monitor LOG]Entering: [RECEIVE]itemRequest(item1)
[Monitor LOG]Entering: [RECEIVE]itemRequest(item2)
[Monitor LOG]Entering: [INVOKE]findItem(item1)
[Monitor LOG]Entering: [INVOKE]findItem(item2)
[Monitor LOG]Entering: [INVOKE]offer(item2)

On receiving the information from the Observer, the **Analyzer** must be able to verify at runtime whether a property has been satisfied or not. Figure 2.a presents the two detailed implement algorithms for the Analyzer. Some variables must be defined first. $I$ is a set of intercepted messages; $P$ is one of the predefined properties which also consists of messages; $P.created$ represents whether the Analyzer is created or not; *Next* function is designed to find the next message in $I$ or $P$; *currentI* and *nextI* represent the current and next message in $I$ respectively; *currentP* represents the current checked message in $P$. The two algorithms will use the four different functions, i.e. *NextCorrect*, *NextFail*, *NextIgnore* and *Constraint* of $P$ defined before. Lines 1-2 initialize *currentP* and *currentI*. Lines 3-5 search the intercepted messages and map them to the first message of $P$. If an intercepted message equals the first message of $P$, a corresponding Analyzer is created, which is shown in lines 6-7. Line 8 visits the next message in $I$. Lines 9-30 deal with the validation process of judging whether or not the intercepted messages satisfy *currentP*. Lines 10-14 mean that the Analyzer will raise an error on condition that *Constraint(currentP)* is not empty and the result of constraint checking is false. In lines 15-29, the Analyzer will check the three next function of *currentP* on condition that *Constraint(currentP)* is empty or the result of constraint checking is true. If *nextI* belongs to *NextCorrect* function of *currentP*, the Analyzer goes to the next message of *currentI* and the next message *of currentP*. If *nextI* belongs to *NextIngore* of *currentP*, the Analyzer just goes to the next message of *currentI*. If *nextI* belongs to *NextError(currentI)*, an error will be raised to explain which message violates $P$. When *currentI* is empty and the next message to be checked in $P$ is required, the Analyzer will raise an error, which is shown in lines 32-35. In other cases, when either *currentP* or *currentI* is empty, the execution will satisfy $P$, which is shown in lines 36-37. Figure 2.b shows the checking constraint algorithm. *Constraint(m)* has time property to show whether it is present, past or future. Lines 1-5 show the present condition, which check whether the messages exchanged at the same time satisfy the constraints. Lines 7-19 deal with the past condition, while lines 8-11 show the condition of Boolean constraints and lines 12-17 show chain constraints. Lines 21-33 deal with future constraints and are also divided into Boolean constraints and chain constraints. We assume that the time complexity of Analyzing algorithm is denoted as $T(A)$ and the time complexity of Checking algorithm is denoted as $T(C)$. The time complexity of two algorithms has the following relationship:$T(A)=O(k)+O(k+j)*T(C)$. Where $k$ is the number of messages in $I$, $j$ is the number of messages in $P$. Concerning Checking algorithm, the worst condition is to judge whether or not the chain event is a substring of $I$. There are existing algorithms, such as KMP algorithm, to solve the problem, and the time complexity is

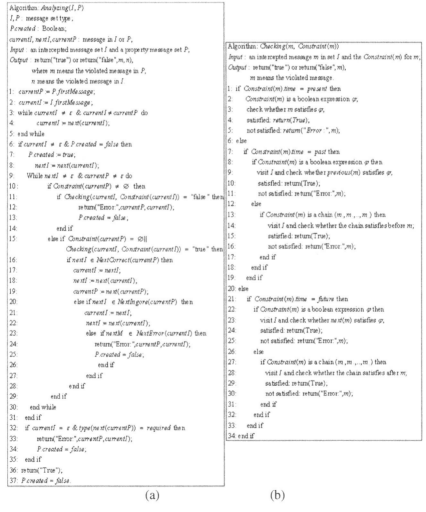

(a)                              (b)

**Fig. 2.** The two detailed implement algorithms of the Analyzer

$O(k+a)$, where $a$ is the maximum number of messages in the chain. So $T(B)=O(b(k+a))$, where $b$ is the number of chains in the property. Therefore, $T(A)=O(k)+O(k+j)*O(b(k+a))$. Generally, $a$ and $b$ is very smaller than $k$, so $T(A)$ is about $O(k(k+j))$.

Sequence1 can be checked against **Prop1** according to our presented approach. The first message in Sequence1 is [RECEIVE]itemRequest which equals the first message of **Prop1**. Thus, a Property Analyzer for **Prop1** is created. Then, the Analyzer intercepts the second message [INVOKE]findItem which belongs to *NextCorrect*([RECEIVE]itemRequest), so this message is true, and the checking process will go to the next step. When message [INVOKE]offerCost is intercepted, it belongs to *NextCorrect*([INVOKE]findItem) and is true also. Then, the next message is [ON MESSAGE]getOrdererData, which belongs to *NextCorrect*([INVOKE]offerCost), and is also true. The last message has a present Boolean constraint, so the Analyzer needs to check whether the current messages satisfy *Constraint*([ON MESSAGE]getOrdererData), and the result is true. Then, there is no next message in **Prop1**. Therefore, **Prop1** has been validated for correctness against the current execution. Finally, the Analyzer sends positive result to users and closes. **Prop2** and **Prop3** can be validated similar to **Prop1**, and results show that they still hold for Sequence1.

Sequence2 is validated against **Prop4**. The first message is [RECEIVE]itemRequest(item1) which equals the first message of **Prop4**, and a Analyzer for **Prop4** is created. When the second message [RECEIVE]itemRequest(item2) is detected, it belongs to *NextCorrect*([RECEIVE]itemRequest(item1)) and is correct. Then [INVOKE]findItem(item1) and [INVOKE]findItem(item2) are ignored because they belong to the *NextIgnore*([RECEIVE]itemRequest(item2)) function. Finally, message [INVOKE]offerCost(item2) is detected, and there is a Past Boolean constraint which needs either [INVOKE]offerCost(item1) or [INVOKE]notAvailable(item1) happens before [INVOKE]offerCost(item2). After checking the messages exchanged, we conclude that it is not the case. So **Prop4** does not hold for Sequence2, and the error information will be notified users of future analyzing.

## 5   Conclusions and Related Works

In this paper, we have presented a more intuitive approach to the problem of monitoring web service compositions described as BPEL processes based on the integration of constraint checking into aspect-oriented programming paradigm.

There are some works related to runtime monitoring of web service compositions in literatures. In [10, 11], an assertion monitor that annotates BPEL codes is proposed. Annotated BPEL processes are automatically translated to monitored processes, and BPEL processes that interleave the business processes with the monitor functionalities. The work in [9] proposes a framework to monitor and validate web service interaction. However, it focuses on web service compositions based on OWL-S and the properties are represented by an extension of Property Specification Pattern (PSP). Barbon et al. in [2] propose an approach sharing us with the same idea to modify ActiveBPEL engine for monitoring purpose. However, they just extend it by adding new modules, and they propose a new property language: RunTime Monitor specification Language (RTML). The work described in [12] provides a framework for monitoring behavioral properties and assumptions expressed in the event calculus. The properties are verified against the collected data by the use of variants of integrity-checking techniques in temporal deductive databases. Banculli et al. in [13] propose an approach to monitor conversational web service compositions based on AOP technique to extend ActiveBPEL engine by the use of algebraic specification to represent properties, and some complex rewrite rules will be used at runtime to validate these properties.

However, note that the potential users of the monitor approach are developers of composite services. These users have no special knowledge of logic or other formalism. It is very difficult for them to manually write the formalism accurately and correctly. Therefore, the main difference to other approaches is that our approach provides a complete graphical front-end for these users that do not have to deal with any particular textual and logical formalism. Moreover, PSC can represent some complex chain properties, directly reason on previous time in web service compositions, which is not easy for other formalism to directly describe.

In the future, we will plan to extend PSC with time constructs to monitor the timed properties and have obtained some primary results[15]. We will also consider extending the framework with techniques that allow for helping users to analyze the result provided by monitors and the automated failure-handling techniques for errors.

# References

1. Papazoglou, M.P., Traverso, P., Dustdar, S., Leymann, F., Krämer, B.J.: Service-oriented computing: A research roadmap. In: Service Oriented Computing (SOC) (2006)
2. Barbon, F., Traverso, P., Pistore, M., Trainotti, M.: Run-time monitoring of instances and classes of Web service Compositions. In: IEEE International Conference on Web Services (ICWS 2006). pp. 63–71 (2006)
3. OASIS. Web Services Business Process Execution Language Version 2.0. Committee Specification (January 2007)
4. Kiczales, G., Lamping, J., Mendhekar, A.: Aspect-oriented programming. In: Akşit, M., Matsuoka, S. (eds.) ECOOP 1997. LNCS, vol. 1241, pp. 220–242. Springer, Heidelberg (1997)
5. ActiveBPEL. The Open source BPEL engine, http://www.activebpel.org
6. Autili, M., Inverardi, P., Pelliccione, P.: A scenario based notation for specifying temporal properties. In: SCESM 2006, pp. 21–28 (2006)
7. Autili, M., Inverardi, P., Pelliccione, P.: Graphical scenarios for specifying temporal properties: an automated approach. Automated Software Engineering 14(3), 293–340 (2007)
8. Kiczales, G., Hilsdale, E., Hugunin, J.: An overview of AspectJ. In: Knudsen, J.L. (ed.) ECOOP 2001. LNCS, vol. 2072, pp. 327–353. Springer, Heidelberg (2001)
9. Li, Z., Jin, Y., Han, J.: A runtime monitoring and validation framework for web service interactions. In: ASWEC 2006: Proceedings of the 17th Australian Software Engineering Conference, pp. 70–79 (2006)
10. Baresi, L., Ghezzi, C., Guinea, S.: Smart monitors for composed services. In: ICSOC 2004, pp. 193–202 (2004)

11. Baresi, L., Guinea, S.: Towards dynamic monitoring of WS-BPEL processes. In: Benatallah, B., Casati, F., Traverso, P. (eds.) ICSOC 2005. LNCS, vol. 3826, pp. 269–282. Springer, Heidelberg (2005)

12. Mahbub, K., Spanoudakis, G.: A framework for requirements monitoring of service based systems. In: ICSOC 2004, pp. 84–93 (2004)

13. Bianculli, D., Ghezzi, C.: Monitoring conversational web services. In: 2nd international workshop on Service oriented software engineering, pp. 15–21 (2007)

14. Muccini, H., Polini, A., Ricci, F., Bertolino, A.: Monitoring architectural properties in dynamic component-based systems. In: Schmidt, H.W., Crnković, I., Heineman, G.T., Stafford, J.A. (eds.) CBSE 2007. LNCS, vol. 4608, pp. 124–139. Springer, Heidelberg (2007)

15. Zhang, P.C., Li, B.X., Sun, M.J., Gong, X.F.: A psc based approach to monitor the timed properties in web service composition. In: 32nd Annual IEEE International Computer Software and Applications Conference, Turku, Finland, IEEE Computer Society, Los Alamitos (2008)

# Efficient Authentication and Authorization Infrastructure for Mobile Users

Zhen Ai Jin[1], Sang-Eon Lee[2], and Kee Young Yoo[1,⋆]

[1] Department of Computer Engineering Graduate School, Kyungpook National University, Daegu, Korea
[2] Department of Information Security, Kyungpook National University, Daegu, Korea
`sarah@infosec.knu.ac.kr, s4ng30n@infosec.knu.ac.kr, yook@knu.ac.kr`

**Abstract.** For providing seamless and secure multimedia service in Wireless network, the protocol needs, efficient design, security and reduction in authentication time when a mobile user move to an other domain (Foreign domain). This paper presents an authentication and authorization infrastructure based on AAA (Authentication, Authorization, and Accounting) to minimize the authentication delay when mobile users move between different domains in wireless networks. In this paper, an AAAH (Home network) and SCC(Security Contest Controller) hierarchy infrastructure were used. A mobile user stores information of SCCs in which the mobile user has been visited. For re-authentication, a mobile user sends the service request and information to the new SCC and it finds the shortest path. Then the new SCC gets security context from the shortest path.

**Keywords:** AAA, SCC, Mobile network, Security, Seamless.

## 1 Introduction

For providing seamless access service, it is a challenge to design, the efficient authentication, authorization and accounting (AAA) protocol for mobile users in a wireless network [1,2]. In AAA, accounting issues are not as time-critical as authentication and authorization, so this paper does not edal with accounting in this paper. The IETF AAA Working Group presented several protocols for a general model: AAA. But these are not applicable to the next generation wireless infrastructure.

In a wireless network, there are potential delays when each time the mobile users get access to a AP(Access Point). Potential delays include three difference kinds of delays:

- First, computational delay that each authentication message requires for generating of keys or the computation of crypt-algorithms.
- Second, media access delay, because of messages sent by other NIC(Network Interface Cards).

---

⋆ Corresponding author.

Y. Ishikawa et al. (Eds.): APWeb 2008 Workshops, LNCS 4977, pp. 155–164, 2008.
© Springer-Verlag Berlin Heidelberg 2008

– Third, authentication exchange delay, because of the long distance between an AP and AAA center.

Computation delay depends on the power station or a variety of crypt algorithms [3], and media access delay depends on the media channel [4]. About authentication exchange delay, it takes 22ms to 25ms to authenticate a single user with a challenge-MD5 algorithm and the RTT(Round Trip Time) is 10ms, it takes 42ms to 45ms for the total authentication procedure for a 4-way handshake [5].

For providing seamless service in a wireless network, each potential delay needs to be reduced. If, the authentication exchange delay occurs a significant amount of delay is due to a proportional long distance.

Usually, the AAA server in the user's home network(AAAH) stores information for verifying the identity of the user. When a user moves from their home network to a foreign network, a user contact the foreign AAA agent(AAAF) and send a service request. To respond to the service request, the AAAF ask the AAAH of the user and the AAAH delivers the challenge information if the user is valid.

For reducing authentication exchange delay, in Kim et'al [5] presented a new AAA authentication mechanism. In this mechanism, the AAAH can delegate an AAA broker to authenticate an mobile user. An AAA broker will be called a security context controller(SCC) hereafter, because the AAA broker stores and controls the security context. It's purpose will be described in detail in the next section.

Most of the proposed authentication protocols for reducing authentication exchange delay either management the SCC to locate appropriate an place or to find the SCC to get some security contexts. But in this paper scheme, the mobile user stores information, which include the SCC ID and date. When the mobile users send the service request, they send the information, too. Then the SCC can find the shortest path which can authenticate the mobile user.

The rest of the paper is organized as follow. Section 2 presents the AAA authentication mechanism and some of the extended AAA authentication mechanism. In section 3 presents a novel authentication architecture. Section 4 analysis the efficient of proposed authentication and authorization architecture by counting the number of the exchange messages. Finally, conclusions are offered we concludes the paper in section 6.

## 2   Related Work

In this section, the AAA authentication mechanism is introduced and the review few authentication mechanisms which have been proposed earlier are reviewed.

### 2.1   Authenticaiton and Authorization Architecture

In Kim et'al [5], proposed, a three-layer AAA authentication mechanism composed of the AAA server, the SCC, and AAA client.

– The AAA server service as an authentication center. It manages the database of all the user's information such as IDs, secret keys and corresponding allowed resources. It generates a set of AV(Authentication Vector) pairs and may authenticate users with them.
– The SCC is like an auxiliary AAA server. It can't generate the AVs, but it can store the AVs which were received from the AAA server. As in Fig.1, the SCC can re-authenticate the mobile user.
– The AAA client is a terminal server such as NAS(Network Access Server) or Access Point router.

In the SCC layer, SCCs can be organized hierarchically as in Figure1. In Kim et'al. [6], researchers presented how to calculate the optimal location of such a SCC in a hierarchically organized network in order to minimize the authentication delay.

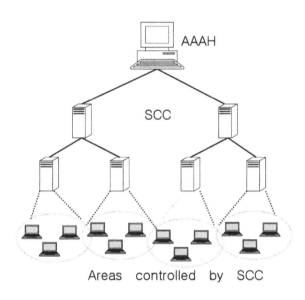

**Fig. 1.** Hierarchy of AAAH and SCCs

The message flow for the authenticaiton of a mobile user using SCCs is described in Fig.2. And the security context information is exchanged only between the SCCs (SCCs include both AAAH and AAA broker).

However in this scheme, when a user moves from the current area to an other area, the new SCC must be determined by the AAAH and the AAAH transfer the security context to a new one. But in high mobility networks, the scheme needs frequent changes of SCC by AAAH, and this procedure accounts for significant delays. The interconnection of the authentication entities is described in Fig.2 .

The protocol issues of security context transfer is already discussed in Georgiades et al. [9] and Wang et al. [8]. In Wang et al. [8], when a mobile user

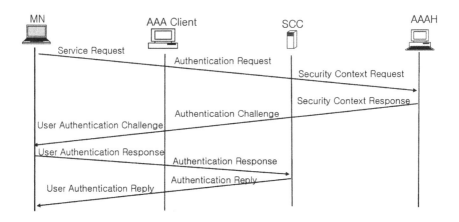

**Fig. 2.** Authentication message exchanges

moves from a domain to a new domain, the new SCC has two options: one is to request the security context for the AAAH. The other is to find the nearest of SCC which still has the security context of the mobile user for authentication. But it generates communication traffic due to the execution of a find algorithm.

In Braun et al. [7], an extension of the concept that the security context can transfer between SSCs without the involvement of the AAAH is proposed. In this scheme, the SCC which a mobile user visited can broadcast the presence of the security context to its neighbors. This broadcast message contains the following information: user_ID, timestamp, SCC_ID, TTL. And when the mobile user move from $SCC_1$ to $SCC_2$, the $SCC_2$ can request to $SCC_1$ and get the security context to easily authenticate the mobile user. This scheme certainly reduces the authentication delay. However this scheme need additional works, e.g SCCs must broadcast message to its neighbors and update it every interval time.

In this paper, there is no message management of SCCs, the mobile node just stores the ID and some information of SCCs which authenticated the mobile user, and this is called a Visted SCC List(VSL). When a mobile user moves domain to a new domain, the new SCC just sends a service request and the VSL. Upon receiving the message, the SCC finds the shortest path to the SCC from which the user can get the security context with the routing tables. So the scheme does not generate communication traffic.

## 3    Efficient Authentication and Authorization Architecture for Wireless Network

In this section, the efficient authentication and authorization architecture is presented to improve the potential delay (The potential delay means the exchange authentication delay). The proposed scheme is also based on peer-to-peer networks. In this scheme, the VSL is added in the service request message and some computation cost for finding the short path using the VSL.

## 3.1   Visited SCC List

In this section, the content of the VSL as shown in Table1 and the algorithm for management as shown Fig.3 are described.

**Table 1.** Notation of VSL

| Notation | Meaning |
|---|---|
| $S_{ID}$ | ID of SCC. |
| $ExpD$ | expiation date of AVs . |
| $N_{AV}$ | Number of remainder AVs which can authenticate a mobile in a SCC. |

**Table 2.** Content of update message

| current $S_{ID}$ | $ExpD$ | $N_{AV}$ | from $S_{ID}$ |
|---|---|---|---|

The content of update message as shown in Table2 is also explained.

Each mobile node has a VSL and update it by two cases. the one is that each node checks the VSL by periods, and if there is SCC with expired date, then the mobile node deletes the SCC from VSL. An other case is when the mobile node receiving the authentication success message which includes the update information from SCC. In this case, if the the $N_{AV}$ of the SCC is zero, then the mobile node also deletes SCC in VSL, but the SCC is the new one, the mobile

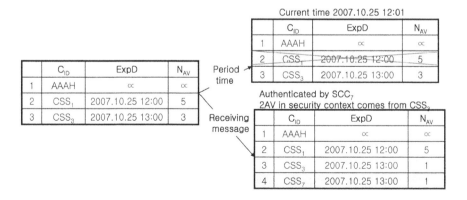

**Fig. 3.** Algorithm of updating VSL table

node inserts the information of the new SCC in to VSL. An example is shown in Fig.3.

In Fig.3, in the first case, the current time exceeded the expiration time of $SCC_1$, the the mobile node deletes $SCC_1$. In the second case, the mobile node added the information of the $SCC_7$ in to VSL.

$$N_{AV_3} = N_{AV_3} - N_{AV_7} - \text{current}(AV) = 3 - 1 - 1 = 1.$$

Then the mobile node updates the $N_{AV}$ of SCC3 as shown in Fig.3.

The advantage of using VSL is that when a SCC receives the VSL from a mobile node, it compares and finds the shortest path which can get the security context.

### 3.2   Proposed Architecture

In this section, we present the whole process of the authentication and authorization scheme are presented as shown Fig.4.

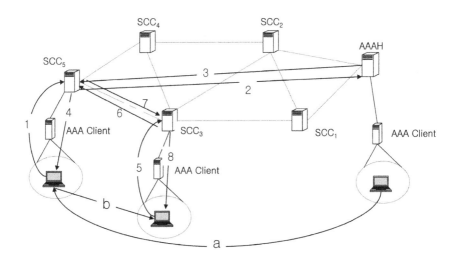

**Fig. 4.** Proposed Authentication Architecture

The "MN ⟶ AAA Client" is a mean mobile node sends a message to the AAA client.

Step 1: MN ⟶ AAA client. A mobile node sends a message which includes a service request and VSL.
Step 2: AAA client ⟶ SCC. AAA client sends the authentication request to SCC. The SCC executes the searching algorithm to find the shortest path SCC.

Step 3: Execute the finding algorithm to find the shortest path. SCC $\longrightarrow$ short-est path SCC, if there is a shortest path SCC in VSL, then sends a security context request to the short path SCC. otherwise SCC $\longrightarrow$ AAAH, sends a security context request to AAAH.

Step 4:  • the shortest path SCC $\longrightarrow$ SCC, if the SCC receiving security con-text request, then the shortest path SCC $\longrightarrow$ SCC, the shortest path SCC sends a part of the security contexts to SCC.
  • otherwise AAAH $\longrightarrow$ SCC, the AAAH sends security context to SCC.

Step 5: SCC $\longrightarrow$ AAA client. The SCC sends authentication challenge and up-date information to AAA client

Step 6: AAA client $\longrightarrow$ Mobile node. AAA client sends a user authentication challenge and update information. Upon receiving the message mobile node updates VSL and sends user authentication response to the AAA client.

Step 7: AAA client $\longrightarrow$ SCC. The AAA client sends the authentication response to SCC.

Step 8: SCC $\longrightarrow$ AAA client. The SCC sends authentication reply to the AAA client.

Step 9: AAA client $\longrightarrow$ Mobile node. The AAA client sends user authentication reply to mobile node. Then the mobile user can provide the service.

In this architecture, when the SCC receives a security context request from a new SCC, it does not transfer completely to the new SCC. It sends a part of the security context and keep some authentication vectors. It might helpful for situations where the mobile node visits the area controlled by the SCC agian. In Fig.4, a path between $SCC_3$ and AAAH is assumed to be longer, then the one between $SCC_3$ and $SCC_5$. In Figure.4, The mobile node moving sequence scenario is from $a$ to $b$ and getting security context sequence scenario is from 1 to 8. A mobile node authenticated by $SCC_5$ first time, and the $SCC_5$ gets the security context from the AAAH. Then the second time, the mobile user visits the area controlled by $SCC_3$. Upon receiving the authentication request and VSL, the $SCC_3$ find the shortest path $SCC_3 - SCC_5$ and $SCC_5$ has the valid security contexts. Then the $SCC_3$ sends security context to $SCC_5$ and $SCC_5$ response to $SCC_3$ with a part of security context.

In this paper, the communication traffic is reduced instead of increasing the message size and some computations in SCC. A mobile node stores the VSL table and manages it. In this paper, computation delay and consumption of some memory is responsible for the reduction of authentication exchange delay [5].

## 4    Evaluation

In this section, an evaluation of novel architecture is given. Usually, the mea-surement of the authentication scheme is the time complexity in the procedure. First, the decrease of communication traffics can be seen. In Fig.5, we the top left node is defined with coordinates(0,0) in 2D coordinate. The previous SCC has coordinates (5,5), and the new SCC has coordinates(4,6). The two SCCs are two hops away from each other, and the AAAH has coordinates(1,2). in Braun et'al. [7], the new SCC meets a node with pointer information for the user's security context and the node has coordinates(3,4). Then the distance between this node and new SCC is 4. The costs for retrieving the security context information for random SCC pairs are $4 * 2 = 8$ and costs for getting the security information context is $2 * 2 = 4$. But in this scheme, the new SCC compares the distances between the new SCC and the previous node - it is 2, and the one between the new SCC and the AAAH - it is 6. Then the new SCC will decide to send the security context request to pervious SCC if it still has the valid security context. In conclusion, this scheme reduces the communication traffic generated by the new SCC to search where the SCC is which can get the security context, while the new SCC executes a searching process with information of VSL. In this example, the scheme reduces the communication cost.

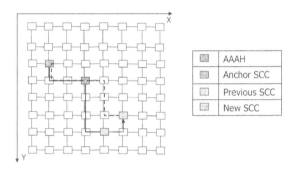

**Fig. 5.** Grid-like organization of SCCs

Subsequently, the total exchange cost for authentication is given in equation (1). The constant number $N$ is the distance between a node and AAAH. Parameter $S$ references the cost for a node, which is authenticated by the SCC in it's controlling area. $H$ is the cost for the new SCC to get a security context from the one hop distance of SCC, then $n$ is the number of hops.

$$C = S + nH, (x < N) \tag{1}$$

In this papers simulation, the AAAH in the middle of a $100 \times 100$ grid of nodes was chosen. The maximum cost of authentication exchange $N$ was set as 210 since the distance between a node and AAAH is 10 hops. Then, the $S$ is 10 and the $H$ is 20 roughly. Fig.6 shows the comparisons with other schemes. And in Figure 6, $|Sreq|$ is the message size of services request size, $|VSL|$ is the message size of VSL information, $n$ is number of mobile nodes and $n_{scc}$ is the number os SCCs.

| | Message size | Communication cost for finding SCC | Storages | |
|---|---|---|---|---|
| | | | SCC | mobile node |
| Georgiades et al [9] | $|Sreq|$ | Yes | 0 | 1 |
| Wang et al [8] | $|Sreq|$ | Yes | $n$ | 1 |
| Braun et al [7] | $|Sreq|$ | Yes | $n$ | 0 |
| Proposed scheme | $|Sreq| + |VSL|$ | No | 0 | $n_{scc}$ |

**Fig. 6.** The total cost of authentication exchange

As in Fig.6, it can be seen that there is no need for finding SCC algorithm in proposed scheme, in other words cost for retrieving security context information is reduced. Then it have an effect on the whole authentication exchange cost significantly. But when the distance is the same with the distance between the new SCC and AAAH, the cost is maximum. Proposed scheme is needed less memory than Braun et al's scheme, because the number of SCCs is small than the number of mobile nodes. A mobile node needs memory space to store VSL information as follow.

$$Total = n \times l(IP) + l(expiretime) + l(number))$$

where, the length of IP is 32bits, the length of expired time is 32bits and length of number is 16bits. It is assumed that the number of nodes which a mobile node visited is 200, then the total memory space is 18,000 (2.250 KB). So, it doesn't hold so much memory. Next, a comparison of the communication overhead of the proposed architecture with other schemes. Service request message size increases by adding VSL information which included the visited SCCs IP. Then, with the same assumption, the length of service request message is increased by as much as 0.8KB, and it can involved the one packet size. It is more efficiency than finding the SCC which holds the security contexts. We expect our scheme can give contributory to improve services quality in wireless networks environment.

## 5   Conclusions

In this paper, an efficient authentication and authorization infrastructure based on AAA used to minimize authentication delays when mobile users move between different domain in wireless networks is presented . An AAAH (Home network) and *SCC* hierarchy infrastructure was also used. Then a mobile user stores the VSL table. For re-authentication in another domain, a mobile user sends the service request and VSL to the new *SCC* and the *SCC* can find shortest *SCC* which the user can get a security context from SCC. The simulation shows the proposed scheme is clearly a reduction in the authentication exchange cost due to finding the shortest path to the SCC.

# Acknowledgements

This research was supported by the MKE( Ministry of Knowledge Economy) of Korea, under the ITRC support program supervised by the IITA (IITA-2008-C1090-0801-0026)

This work was supported by grant No. R01-2007-000-10614-0 from the Basic Research Program of the Korea Science and Engineering Foundation.

# References

1. Rensing, C., Hasan, Karsten, M., Stiller, B.: A Survey and a Policy-Based Architecture and Framwork. IEEE Newtwork, 22–27 (November/December 2002)
2. Mitton, D., St. Johns, M., Barkley, S., Nelson, D., Patil, B., Stevens, M., Wolff, B.: Authentication, Authorization, and Accouning: Protocol Evaluation. RFC 3127 (June 2001)
3. Stallings, W.: Cryptography and Network Security: Principles and Practice. Prentice-Hall, Englewood Cliffs (1999)
4. Cam-Winget, N., Smith, D., Amann, K.: Proposed new AKM for Fast Roaming (January 2003) doc.:IEEE 802.11-03/008r0
5. Kim, H., Afifi, H.: Improving Mobile Authentication with New AAA Protocols. In: IEEE International Conference on Communications 2003 (ICC 2003) (May 2003)
6. Kim, H., Ben-Ameur, W., Afifi, H.: Toward Efficient Mobile Authentication in Wireless Inter-Domain. In: 3rd Workshop on Applicaitons and Services in Wireless Networks(ASWN), Bern, Switzerland (July 2003)
7. Braun, T., Kim, H.: Efficient Authentication and Authorization of Mobile Users Based on Peer-to-Peer Network Mechanisms. In: Proc. of Hawaii Int. Conf. on System Sciences (HICSS), January 30, 2005, pp. 306b–306b. IEEE, Los Alamitos (2005)
8. Georgiades, M., Akhtar, N., Ploitis, C., Tafaziolli, R.: AAA Context Transfer for Seamless and Secure Multimedia Services over ALL-IP Infrastructures. In: 5th European Wireless Conference(EW 2004), Barcelona, February 24-27 (2004)
9. Wang, H., Prasad, A.: Security Context Transfer in Vertical Handover. In: 14th IEEE 2003 International Symposium on Personal, Indoor and Mobile Radio Communication Processing, Beijing, September 7-10 (2003)

# An Effective Feature Selection Method Using the Contribution Likelihood Ratio of Attributes for Classification

Zhiwang Zhang[1], Yong Shi[2], Guangxia Gao[3], and Yaohui Chai[4]

[1] School of Information of Graduate University of Chinese Academy of Sciences, China
Chinese Academy of Sciences Research Center on Fictitious Economy and Data Science,
Beijing (100080), China
zzwmis@163.com
[2] Research Center on Fictitious Economy and Data Science, Chinese Academy of Sciences,
Beijing 100080, China; College of Information Science and Technology,
University of Nebraska at Omaha, Omaha NE 68182, USA
yshi@gucas.ac.cn
[3] Foreign Language Department, Shandong Institute of Business and Technology, Yantai,
Shandong 264005, China
Gaoguangxia2006@126.com
[4] School of Management of Graduate University of Chinese Academy of Sciences, China
Chinese Academy of Sciences Research Center on Fictitious Economy and Data Science,
Beijing (100080), China
jinhonghui@163.com

**Abstract.** Feature selection is a very crucial step in data mining process. It aims to find the most important feature subset from a given feature set without degradation of classifying information. As for the traditional feature selection method, the number of candidate feature subsets created by algorithm in an iterative computational way is exponential in the size of the initial attribute set. And relevant algorithm occupies a lot of the system resources in time and space. In this paper, we study and develop a novel feature selection method and provide its mathematic principle, which is based on the factors of attributes contributing to target attribute and their maximum information divergence value (MIDV) to select small enough feature subset and improve the classification accuracy. And then the extensive experiment shows that our proposed method is very efficient in computational performance and scalability than traditional methods.

**Keywords:** Feature Selection, Contribution Factors, Maximum Information Divergence Value.

## 1 Introduction

With the developments of information technologies, all kind of systems produce and generate a large number of structured and semi-structured data and information. Such as text files, web files, and transaction data files so on. When dealing with large data sets, it is often the case that the information available is redundant for data mining

Y. Ishikawa et al. (Eds.): APWeb 2008 Workshops, LNCS 4977, pp. 165–171, 2008.
© Springer-Verlag Berlin Heidelberg 2008

application to some extent. Consequently, the reduction of the original feature set to a smaller one preserving the relevant information while discarding the redundant information is referred to as feature selection [Vanda and Giovanni, 2006].

Applying feature selection to data mining application may mainly bring benefits in the following: a) Reducing the amount of information needed by data mining algorithm and computational time and space complexities. b) Learning and gaining compact and high quality rules from data. c) Easily and understandable expressing knowledge and decision rules got from the algorithms.

This paper is organized as follows. In the section of the flow of feature selection, we introduction the four fundamental steps based on the experience of predecessors. And then we present the basic principles and deduce process of feature selection using the contribution factors likelihood ratio. Subsequently, the section of algorithm gives inputs, outputs, the flow and steps of the algorithm. Afterwards we bring out the experiments processing and results of the algorithm which can be applied to different data sets. Finally, we draw the conclusions and future works of further improvements about the feature selection method.

## 2   The Flow of Feature Selection

According to the work of [Langley, 1994], feature selection method is based on four main steps as following Fig.1. They are: 1) Generating candidate feature set. 2) Computing evaluation function. 3) Stopping criterion. 4) Evaluating selected feature subset.

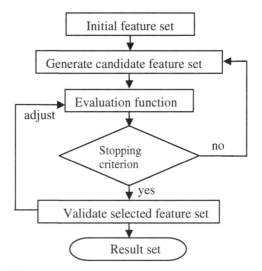

**Fig. 1.** Four main steps of feature selection method

Firstly, generating candidate feature set is the procedure that produces candidate subset from initial feature set. Theoretically the number of candidate subset from a set of N is $2^N$. Therefore, it is very crucial that choosing a proper candidate will directly affects the efficiency of feature selection. As for selection direction, we have three methods: one is forward method that the generation starts with an empty set, and then forming a novel candidate feature set by means of adding a new feature to result set at each iteration. By contrast, the other one is backward method that begins with the whole of feature set, and then at each step removes one which is not satisfied with predefined conditions. Alternatively, the hybrid method that is producing candidate feature set in forward, backward or random way.

Then evaluation function is used to measure the quality of candidate feature subset, that is to say, to evaluate the classification performance of the candidate subset. According to different measurement method, evaluation function may be classified into four types: 1) Distance measurement, computing the distance among different feature for classification power. 2) Information measurement, calculating the quantity of information retained by given feature set. 3) Correlation measurement, indicating the capacity of the selected features to predict the value of others features. 4) Consistency measurement, evaluating the capacity of the selected features to separate the objects in different classes.

The stopping criterion is a judgment that the selected feature is satisfied with predefined criterion which avoid algorithm occupying too much system resources.

Finally, the evaluation of selected features subset is to measure the quality of feature selection method by means of running data mining algorithm on these feature subset.

## 3   Feature Selection Using the Contribution Likelihood Ratio of Features

For a given training set $S = \{(x_1, \omega_1), \cdots, (x_l, \omega_l)\}$, thereinto $x_i$ is points which are from set $X \subseteq R^n$, at the same time, classification label $\omega_i$ come from set $Y \subseteq R^n$. In addition, we use $x = (x_1, x_2, \cdots, x_d)$ to express d-dimensions input vectors.

Here, we define a statistic measurement which compute the metric value of the attribute variable $x_k$ $(k = 1, 2, \cdots, n)$ contributions to the objective variable $\omega_i$, It is contributive factor: $c_{ki} = p(x_k, \omega_i) / p(x_k) p(\omega_i)$.

It is needed to notice that $c_{ki}$ means a certain extent $x_k$ helps to $\omega_i$. And we know the value of $c_{ki}$ is 1 if $x_k$ is independent of $\omega_i$, that is to say, $p(x_k, \omega_i) = p(x_k) p(\omega_i)$.

Thus, we get the result of $c_{ki} = 1$. The more contributive value, the more distance $c_{ki}$ is far away from constant 1.

For binary classification, similarly, we may get contributive factor $c_{kj}$ which show boosting function of the attribute variable $x_k$ to the objective variable $\omega_j$. Namely, $c_{kj} = p(x_k, \omega_j) / p(x_k) p(\omega_j)$.

Then we may define the likelihood ratio $c_{ij}$ as the ratio of contribution factors according to $c_{ki}$ and $c_{kj}$ of attribute $x_k$:

$$c_{ij} = \ln(c_{ki}/c_{kj}) = \ln(p(x_k, \omega_i)/p(x_k, \omega_j)) - \ln(p(\omega_i)/p(\omega_j)).$$

And then average classifiable information $c_{ij}$ of class $\omega_i$ to class $\omega_j$ is: $Cij = E(c_{ij})$,

For continuous variable, we have:

$$Cij = E(c_{ij}) = \int_X p(x_k/\omega_i)[\ln(p(x_k,\omega_i)/p(x_k,\omega_j)) - \ln(\omega_i/\omega_j)]dx,$$

Note that $p(x_k/\omega_i)$ is attribute $x_k$'s density function, we may extend it to combine density function of multi-attributes.

Besides, for discrete variable, we have to transform computational method of $c_{ij}$:

$$Cij = E(c_{ij}) = \sum_{k=1}^{n_i} (p(x_k,\omega_i)/p(\omega_i))[\ln(p(x_k,\omega_i)/p(x_k,\omega_j)) - \ln(\omega_i/\omega_j)].$$

Similarly, average classifiable information $c_{ji}$ of class $\omega_j$ to class $\omega_i$ is:

$$Cji = E(c_{ji}) = \int_X p(x_k/\omega_j)[\ln(p(x_k,\omega_j)/p(x_k,\omega_i)) - \ln(\omega_j/\omega_i)]dx.$$

In the end, the information divergence value (IDV) $IDV_d$ about class $\omega_i$ and $\omega_j$ on attribute $x_k$ is:

$$IDV_d = Cij + Cji = \int_X [p(x_k/\omega_i) - p(x_k/\omega_j)][\ln(p(x_k,\omega_i)/p(x_k,\omega_j)) - \ln(\omega_i/\omega_j)]dx,$$

Here d is the number of attribute or dimension of data set.

## 4  Feature Selection Algorithm

In this part, we give the algorithm which implement feature selection using the likelihood ratio of contribution factors (CFLRA), the following is the description:

***Algorithm:*** CFLRA

***Input:*** the attribute set $x = (x_1, x_2, \cdots, x_D)$, iif. $D$ is maximum dimension number, the convergence threshold of MIDV is $\varepsilon(0 < \varepsilon < 1)$.

***Output:*** the minimum attribute set $x^* = (x_1, x_2, \cdots, x_d)$, iif. $x^* \subseteq x$ and $d \le D$.

***Algorithm flow:*** This algorithm mainly includes the following five steps:

① Initialize parameter $d = 1$, and the objective attribute set $x^* = \{\}$;

② According to the information divergence value $IDV_d$'s computational method above, we compute and get many information divergence value $IDV_{di}$ $(i = 1, 2, \cdots, D)$ of the different attribute $x_i$ on the condition that dimension is 1, that is to say, subscript $d = 1$.

③ Compute the maximum of the information divergence values(MIDV):

$k = \max(IDV_{di} \mid 1 \le i \le D)$, and we get new selected attribute set: $x^* = x^* \cup \{x_k\}$.

In addition, we need to compute adjoin the two maximum information divergence values $IDV_{dk_1}$ and $IDV_{dk_2}$:

If $\left| IDV_{dk_1} - IDV_{dk_2} \right| < \varepsilon$ $(k_1 = i, k_2 = i+1, 1 \le i \le D)$, then the flow turn to step ⑤.

In the other words, if $d = D$, then the flow turn to step ⑤ too.

④ Then we increase dimension by 1: $d = d+1$, and we repeat step ②, at the same time, subscript $i$ satisfy with the condition: $i \ne k$.

⑤ Exit the process of computing the information divergence values, finally we get reduced and target attribute set $x^*$.

## 5  Experiments and Results

In this section, we evaluate our method using data set mushroom which sources from mushroom database donated by Jeffrey Schlimmer. This data set includes descriptions of hypothetical samples corresponding to 23 species of gilled mushrooms in the Agaricus and Lepiota Family. Each species is identified as definitely edible, definitely poisonous, or of unknown edibility and not recommended. At the same time, the data set has 8124 instances or observations and 22 attributes which include mainly cap, odor, gill, stalk, veil, ring, spore, population, and habitat etc. Consequently, the attribute set structure and the distribution of values are showed in the following Tab.1.

**Table 1.** The attribute set and the values distribution of mushroom

| Attribute | Values |
|---|---|
| cap-shape | bell, conical, convex, flat, knobbed, sunken |
| cap-surface | fibrous, grooves, scaly, smooth |
| cap-color | brown, buff, cinnamon, gray, green, pink, purple, red, white, yellow |
| bruises | bruises, no |
| odor | almond, anise, creosote, fishy, foul, musty, none, pungent, spicy |
| gill-attachment | attached, descending, free, notched |
| gill-spacing | close, crowded, distant |
| gill-size | broad, narrow |
| gill-color | black, brown, buff, chocolate, gray, green, orange, pink, purple, red, white, yellow |
| stalk-shape | enlarging, tapering |
| stalk-root | bulbous, club, cup, equal, rhizomorphs, rooted, missing |
| stalk-surface-above-ring | ibrous, scaly, silky, smooth |
| stalk-surface-below-ring | ibrous, scaly, silky, smooth |
| stalk-color-above-ring | brown, buff, cinnamon, gray, orange, pink, red, white, yellow |
| stalk-color-below-ring | brown, buff, cinnamon, gray, orange, pink, red, white, yellow |
| veil-type | partial, universal |
| veil-color | brown, orange, white, yellow |
| ring-number | none, one, two |
| ring-type | cobwebby, evanescent, flaring, large, none, pendant, sheathing, zone |
| spore-print-color | black, brown, buff, chocolate, green, orange, purple, white, yellow |
| population | abundant, clustered, numerous, scattered, several, solitary |
| habitat | grasses, leaves, meadows, paths, urban, waste, woods |

For the sake of simplification, we use binary classification target, that is to say, target attribute has two different value of edible and poisonous mushroom. And that its class distribution is 51.8% edible parts and 48.2% poisonous parts.

In the experiment, firstly we choose the parameter $\varepsilon = 0.01$, then compute the information divergence values and obtain its maximum on the condition of different dimension in iterative way. In the first iteration, that is to say, $d = 1$, the MIDV is 12.6017 and the selected attribute is spore-print-color correspondingly. In the second loop, added 1 to the dimension $d$, the MIDV is 38.1027 and the selected attribute is gill-color. Similarly, the MIDV is 63.6704 and the selected attribute is cap-shape in the third iteration. And then the MIDV is 72.1041 and the selected attribute is gill-spacing in the fourth iteration. And the MIDV is 72.4590 and the selected attribute is population in the fifth iteration. And the MIDV is 73.3492 and the selected attribute is gill-attachment in the sixth iteration. Finally, in the seventh loop, the MIDV converge at 73.3492 and the selected attribute are veil-type and veil-color, here the amount that remains after the MIDV which is subtracting from previous one is less than predefined threshold $\varepsilon$, so the algorithm flow is finished at this step.

The results of the experiments may express as Fig. 2. in the following:

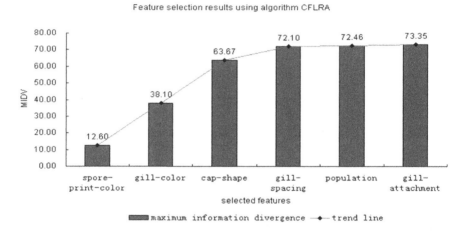

**Fig. 2.** Feature selection results using algorithm CFLRA

In addition, we also got good effect on both performance and accuracy of classification where the selected feature set is produced from the algorithm CFLRA in the different data mining applications. However, for continuous variables we firstly need to discretize them, and then use algorithm CFLRA to select feature subset and further modeling analysis.

## 6   Conclusions and Future Work

The results obtaining from the experiments in this paper show that the feature selection algorithm using the likelihood ratio of contribution factors can efficiently get solution and evidently improve the performance of classification. In general, we begin with the whole of feature set and use forward method to generate candidate feature subset. Besides, it takes not only the correlation of different attributes and target attribute but also the contribution of attributes versus target into account. Additionally, the information divergence value is used as evaluation criterion which indicates the classification power of different feature. In conclusion, algorithm CFLRA is alternatively a good feature selection method in data mining application. However, we need to improve the algorithm further so that it could be applied to regression, association, clustering problems in the future.

**Acknowledgements.** This research has been partially supported by a grant from National Natural Science Foundation of China (#70621001, #70531040, #70501030, #70472074), 973 Project #2004CB720103, Ministry of Science and Technology, China, and BHP Billiton Co., Australia.

# References

1. Mitchell, T.: Machine Learning. McGraw-Hill, New York (1997)
2. de Angelis, V., Felici, G., Mancinelli, G.: Feature selection for data mining. In: Triantaphyllou, E., Felici, G., (eds.): Data Mining and Knowledge Discovery Approaches based on Rule Induction Techniques. Massive Computing Series, vol. 6, pp. 227-252. Springer, Heidelberg (2006)
3. Langley, P.: Selection of relevant features in machine learning. In: Proceedings of the AAAI Fall Symposium on Relevance. AAAI Press, Menlo Park (1994)
4. Duda, R.O., Hart, P.E.: Pattern Classification, 2nd edn., pp. 90–101. Elsevier Science, USA (2003)
5. Fukunaga, Keinosuke: Introduction to Statistical Pattern Recognition, 2nd edn., pp. 489–503. Elsevier Academic Press (1999)
6. Ruiz, R., Aguilar–Ruiz, J.S.: Analysis of Feature Rankings for Classification, Spain
7. Abe, N., Kudo, M.: Entropy Criterion for Classifier-Independent Feature Selection, Japan
8. Abe, N., Kudo, M.: A Divergence Criterion for Classifier-Independent Feature Selection, Japan
9. Cang, S., Partridge, D.: Feature ranking and best feature subset using mutual information. Springer, London (2004)
10. Chizi, B., Maimon, O.: Data Mining & Knowledge Discovery Handbook. Springer Science, pp. 93–109 (2005)
11. Ian, H., Witten, I.H., Frank, E.: Data Mining: Practical Machine Learning Tools And Techniques, 2nd edn. Elsevier Inc., Amsterdam (2005)
12. Han, J., Kamber, M.: Data Mining: Concepts and Techniques, pp. 10–19. Morgan Kaufmann Publishers, San Francisco (2001)
13. Ye, N.: The Handbook of Data Mining, pp. 414–417. Lawrence Erlbaum Associates, Mahwah (2003)
14. Nixon, M.S., Aguado, A.S.: Feature Extraction and Image Processing. Elsevier Newnes, Amsterdam (2002)
15. Fukunaga, K.: Introduction to Statistical Pattern Recognition, 2nd edn., pp. 489–503. Elsevier Academic Press, Amsterdam (1999)
16. Marques de Sa, J.P.: Pattern Recognition Concepts Methods and Applications, pp. 65–69. Springer, Heidelberg (2001)

# Unsupervised Text Learning Based on Context Mixture Model with Dirichlet Prior

Dongling Chen[1,2], Daling Wang[1], and Ge Yu[1]

[1] Northeastern University, Shenyang 110004, P.R. China
[2] School of Information, Shenyang University, Shenyang 110044, P.R. China
Chen.dongling@yahoo.com.cn,{dlwang,yuge}@mail.neu.edu cn

**Abstract.** In this paper, we proposed a bayesian mixture model, in which introduce a context variable, which has Dirichlet prior, in a bayesian framework to model text multiple topics and then clustering. It is a novel unsupervised text learning algorithm to cluster large-scale web data. In addition, parameters estimation we adopt Maximum Likelihood (ML) and EM algorithm to estimate the model parameters, and employed BIC principle to determine the number of clusters. Experimental results show that method we proposed distinctly outperformed baseline algorithms.

**Keywords:** Dirichlet prior, Finite mixture model, Text clustering, Bayesian, Parameter estimation.

## 1 Problem Description

More and more information spring up on internet, such as Internet news feeds, electronic mail, blogs, medical patient records, and so on. It is very important to cluster text document automatically and unsupervised in personalized search. By far, there exist two kinds of different cluster approaches: first, agglomerative approach, which build up clusters by iteratively sticking similar things together; second, mixture model approach, which learn a generative model over the data, treating the classes as hidden variables. For two approaches, people are more prone to mixture model approach, for two reasons: agglomerative approach must be in ad hoc manner, which is intractable for large data sets; and no sound probabilistic foundation.

Existing statistical text clustering algorithms based on mixture model can be divided into two groups: supervised [1,2] and unsupervised [3,4,5,6,7,8]. In the supervised classification framework, we have labeled examples; however, in many real life problems it is very difficult or impossible to get examples from one or more classes. The wide coverage of topics, dynamics of the internet information makes it extremely difficult to classify large-scale web information. The lack of examples from some of the classes makes it unfeasible to apply on large-scale web data clustering. Let y denote a discrete dependent, outcome, target output variable, and $z$ a vector of independent, input, predictor, or attribute variables, and classification involves predicting the discrete outcome variable y as accurate as possible using the information on the z variables.

Y. Ishikawa et al. (Eds.): APWeb 2008 Workshops, LNCS 4977, pp. 172–181, 2008.

$$P(y, z) = P(y)P(z \mid y) = P(y)\sum_x P(x \mid y)P(z \mid y, x) \tag{1}$$

In accordance with this basic form, there are many mixture probability models, such as K-means, Gaussian mixtures, Naïve Bayesian, and so on. Those approaches have been widely used in various fields. Nevertheless, those models are not the best choice in all applications, they are often fail to discover true structure, especially, when the input variables have multiple probability distributions. For example, if a text may be contain multiple topics, those approach mentioned above cannot suitable for multiple class classification. Therefore, in this paper we will show that we introduce context variable which have Dirichlet prior distribution in bayesian mixture model, which can be a very good choice to overcome the disadvantages above mentioned. The Dirichlet distribution as prior mixture model is the multivariate generalization of the Beta distribution, which offers considerable flexibility and easy to use.

The rest of the paper is organized as follows. In Section 2, we list existing mixture model applications, in section 4, we gives the model selection approach; The experimental setup and evaluation are described in Section 5. Finally, in Section 6, presents our conclusions and future works.

## 2   Existing Mixture Model Applications

Mixture models form one of the most widely used classes of generative models for describing structured and clustered data [9]. A vast amount of mixture models be proposed recently, most of it is based on local search techniques, such as Naïve-Bayes (NB) methods [10,11]. The NB assumes mutual independence of the z variables within levels of y, $p(z \mid y) = \prod_l P(z_l \mid y)$ .while the exact form of the conditional density $P(z_l \mid y)$ depends on the scale type of $z_l$ .less restricted forms for $p(z \mid y)$ are used in Bayesian tree classifiers and in discriminant analysis.

Recently, Kleinberg and Sandler[4,5] have shown that there is a combinatorial algorithm, reconstruct the underlying term distributions for each document .then given the topical distributions, the algorithm reconstruct accurately the relevance to each of the topics for each document. Another classical probabilistic generative models are PLSA model, which is proposed by Hofmann [12], which models a document as a mixture of aspects, where each aspect is represented by a multinomial distribution over the whole vocabulary. Thus, each word is generated from a single topic, and different words in a document may be generated from different topics.

Nowadays, some extensions of mixture models are explored, for example, Blei proposed LDA model to overcome those shortcomings. LDA[6] can avoid overfitting in PLSA and extract a set of themes from a document collection. A document is considered as a mixture of topics. Liu et.al.[13] and Li and Yamanishi[14] also proposed methods for topic analysis using a finite mixture model. Especially, Liu et al. considered the problem of selecting the optimal number of mixture components in the context of text clustering; Author-Topic(AT) model is proposed in [15], AT model is a similar Bayesian network, in which each authors' interests are modeled with a mixture of topics, and so on.

We have reviewed so many clustering approaches based on mixture model. These methods have been proven be efficient from their experimental results. However, these methods are suffer from the deficiency that they cannot grasp text contextual information, such as time, location at which the document was produced, the author who wrote the document and its publisher. According contents of text document with the same or similar context are often correlated in some way. This is core idea of this paper.

## 3   Problem Formulation

### 3.1   The Finite Dirichlet Mixture Model Formal Description

First, we give bayesian finite mixture model, $X = X_1,..., X_n$ denote m discrete random variables, $\vec{d}$ is a vector, $\vec{d} = (X_1 = x_1,...X_m = x_m)$, where $x_i \in \{x_{i1},...x_{in_i}\}$, a random sample $D = (\vec{d_1},...,\vec{d_N})$ is a set of Ni.i.d. data instantiations, then,

$$p(\vec{d}) = \sum_{k=1}^{K} P(Y = y_k)P(\vec{d} \mid Y = y_k) \tag{2}$$

Where the value of the discrete clustering random variable Y correspond to the separate clusters of the instantiation space, and each mixture distribution $P(\vec{d} \mid Y = y_k)$ models one data producing mechanism. It follows that we can assume the variables Xi inside each cluster to be independent and thus (1) becomes

$$p(\vec{d}) = P(X_1 = x_1,..., X_m = x_m) = \sum_{k=1}^{K} P(Y = y_k)\prod_{i=1}^{m} P(X_i = x_i \mid Y = y_k) \tag{3}$$

Where J denotes latent numbers of clusters; $\pi_j$ denotes the proportion of each cluster in the whole components data, where $\pi_j > 0$, $\sum_{j=1}^{J} \pi_j = 1$, and the probability density of k-dimensional Dirichlet distribution $D(\alpha_{j1},...,\alpha_{jk})$ is[16]:

$$f(d;V_j) = \frac{\Gamma\left[\sum_{s=1}^{k} \alpha_{js}\right]}{\prod_{s=1}^{k} \Gamma(\alpha_{js})} \prod_{s=1}^{k} d_s^{\alpha_{js}-1} \tag{4}$$

Here, $x = (x_1,...x_k)$ is k-dimensional vector, the restriction is $\sum_{s=1}^{k} x_s = 1$, and $V_j$ denotes parameters $(\alpha_{j1},...,\alpha_{jk})$, $\alpha_{js} > 0$, $s = 1,...,k$, V can be characterized as $V = (\pi_1,...\pi_J, V_1,...V_J)$, V,J are all unknown parameters. If the document that can be observed are noted down $d_i = (d_{i1},...,d_{ik})$, $i = 1,...,n$, then given the V,J the joint probability density can be denoted as follows:

$$p(D \mid V, J) = \prod_{i=1}^{n} \sum_{j=1}^{J} \pi_j f(d_i;V_j) \tag{5}$$

If we introduce a virtual variable for each observed document data $z_i = (z_{i1},...,z_{iJ})$ , $i = 1,...,n$ , here, $z_{ij} = \begin{cases} 1 & if\ d_i\ come\ from\ f(d;V_j) \\ 0 & otherwise \end{cases}$ , then characterized $Z=\{z_i, i=1,...,n\}$, then the joint probability is:

$$p(D,Z|V,J) = \prod_{i=1}^{n}\prod_{j=1}^{J}[\pi_j f(d_i;V_j)]z^{ij} \tag{6}$$

## 3.2 Defining Model Priors

If a set of optimum parameters $(V^*, J^*)$ that can make equation (7) get its maximum can be found, we think the most optimum model to fit the document dataset could be found. In order to get a set of optimum $(V^*, J^*)$, we want to introduce an appropriate prior knowledge for parameters. This knowledge may be obtained from domain specific lexicons, or training data in this domain as in [17]. However, it is impossible to have such knowledge for a huge and random dataset. So, in this section, we would adopt Bayesian method to give different parameters a prior knowledge.

For parameter $J$, its prior knowledge $P(J)$ is a discrete probability distribution on natural number. So, $P(J)$ maybe a discrete even distribution on $\{1, 2,..., n\}$ as no information prior. Indeed, given an observed dataset, a user often has some knowledge about what aspects or how many aspects in it.

For parameter $V$, its prior knowledge $P(V|J)$, if given $J$, then $(\pi_1,...,\pi_J)$ and $(V_1,...,V_J)$ are conditional independence each other, namely:

$$P(V|J) = P(\pi_1,...,\pi_J|J) \cdot P(V_1,...,V_J|J) \tag{7}$$

The prior knowledge of $(\pi_1,...\pi_J)$ usually can be get through $D(\beta_1,...,\beta_J)$, here, $\beta_j > 0$ , $j = 1,...,J$ is hyper-parameter. We denote $\beta = \sum_{j=1}^{J}\beta_j$ , and according to Jeffery rule, take $\beta_1 = ... = \beta_J = \frac{1}{J}$ , so in this paper, we assume that $P(V_1,...V_J) \propto 1$.

## 3.3 Parameters Estimation

Estimating the parameters, we can judge how many components in the mixture model, i.e., how many clusters can be generative. Usually Maximum Likelihood (ML) is most popular approach. The likelihood for the complete data set:

$$P(X|\pi_j) = Multinomial(.|\pi_j) = Dir(.|\pi_1^*,...\pi_n^*) \tag{8}$$

Calculate the $\max_{\pi_j} P_{\pi_j}(X|\pi_j)$ with the constraints: $\pi_j > 0$ $\sum_{j=1}^{J}\pi_j = 1$ , the graphical representation as follows:

Then we will use Expectation Maximization (EM) estimating MLE. With the prior defined above, in the next step, we have three tasks: 1) to judge the numbers of clusters. 2) Given $J$, to estimate the proportions, as well as cluster feature parameters

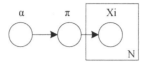

**Fig. 1.** Graphical Representation of multinomial sampling with Dirichlet prior

V. 3) Given $J$, to cluster document set D, that is to say, to estimate variational parameter Z.

According to Bayes formula,

$$P(V \mid D, J) \propto P(D \mid V, J) P(V \mid J) \tag{9}$$

It is very difficult to calculate $P(D \mid V, J)$, because $P(D \mid V, J)$ is a mixture model, so incorporate section 3.2, we introduce a virtual variable Z, the equation (12) can be rewrited as follows:

$$P(V \mid D, J) = \frac{P(V, Z \mid D, J)}{P(Z \mid D, V, J)} \tag{10}$$

Where  $P(V, Z \mid D, J) \propto P(D, Z \mid V, J) P(V \mid J)$

According to equation (9)(10), those parameters are easily be computed, in the following section, we can calculate $V^*$ based on EM algorithm. The EM algorithm is commonly applied to find a local maximum-likelihood estimate of the parameters in situations when the observable data is incomplete and the model depends on unobserved latent variables.

E-step:

$$Q(V, V^{(t)}) = E_{(Z \mid D, V^{(t)}, J)} \{\log P(D, Z \mid V, J)\} + \log P(V \mid J) \tag{11}$$

Further unfolded computation as follows:

$$Q(V, V^{(t)}) = \sum_{i=1}^{n} \sum_{j=1}^{J} \frac{\pi_j f(x_i; V_j^{(t)})}{\sum_{j=1}^{J} \pi_j f(x_i; V_j^{(t)})} \times$$

$$\{\log \pi_j + \log \Gamma[\sum_s \alpha_{js}] + \sum_s (\alpha_{js} - 1) \log x_{is} - \sum_s \log \Gamma(\alpha_{js})\} + \sum_{j=1}^{J} (\beta_j - 1) \log \pi_j \tag{12}$$

M-step: To calculate the $Q(V, V^{(t)})$ maximum, get $V^{(t+1)}$. According to Lagrange it is easily to known:

$$\pi_j^{(t+1)} = [\sum_{i=1}^{n} \frac{\pi_j f(x_i; V_j^{(t)})}{\sum_{j=1}^{J} \pi_j f(x_i; V_j^{(t)})} + \beta_j - 1] \Big/ (n + \beta - J) \tag{13}$$

Whereas $\alpha_{jh}^{(t+1)}$ is equational solution of (18)

$$\sum_{i=1}^{n} \frac{\pi_j f(x_i; V_j^{(t)})}{\sum_{j=1}^{J} \pi_j f(x_i; V_j^{(t)})} [\frac{\Gamma'[\sum_s \alpha_{js}]}{\Gamma[\sum_s \alpha_{js}]} + \log x_{ih} - \frac{\Gamma'(\alpha_{jh})}{\Gamma(\alpha_{jh})}] = 0 \qquad (14)$$

Iterating starts from t=0, between the E-step and M-step until the monotonically increasing log-likelihood function gets a local optimal limit, that is to say, the difference values between $V^{(t+1)}$ and $V^{(t)}$ will be stabilized and little enough, we get the convergence point is a local maximum point of $Q$.

# 4   Model Selecting

According to Bayesian method, we can cluster document set $D$ through posterior distribution $P(Z \mid D, J)$, we may use the MAP estimator [18], that is:

$$\hat{Z} = \arg \max P(Z \mid D, J)$$

Namely,

$$P(Z \mid D, V^*, J) = \frac{P(D, Z \mid V^*, J)}{P(D \mid V^*, J)} = \prod_{i=1}^{n} \prod_{j=1}^{J} [\frac{\pi_j^* f(x_i; V_j^*)}{\sum_{j=1}^{J} \pi_j^* f(x_i; V_j^*)}]^{z_{ij}} \qquad (15)$$

According to Equation (21), we can make sure what cluster document $d_i$ belongs to. Another core problem is to judge the numbers of clusters, under the Bayes analysis framework, we adopt BIC rule to make sure the numbers of clusters. The detailed computation process as follows [16,19,18]:

$$BIC_J = 2 \log P(D \mid V^*, J) - m_J \log(n) \approx 2 \log(P(D \mid J)) \qquad (16)$$

Where $m_J = (J - 1) + Jk$ , $J$-1 is the ponderance numbers of $\pi_1, ..., \pi_J$ .

# 5   Experiments and Evaluations

In this section, we present some experiment results of we proposed clustering algorithm based on Dirichlet prior mixture model and model parameter estimation. We evaluate the effectiveness of our model on famous 20-Newsgroup dataset. Experiments include two parts: the first, to evaluate clustering algorithm performance; the second, we make contrast experiments with baselines.

We implemented all algorithms in JAVA and all experiments have been executed on Pentium IV 2.4GHz machines with 1GB DDR main memory and windows XP as operating system.

Notice, our clustering algorithm which is actually a Mixture Model with Dirichlet prior, so, in the following, we called it as MMDB.

## 5.1   Dataset

we conducted experiments on the publicly available 20-Newsgroups collections. This data set contains text messages from 20 different Usenet newsgroups, with 1000 messages collected from each newsgroup. We derived three group datasets just as [20,21]. The first group, News-Similar-3 consists of messages from 3 similar newsgroups (comp.graphics, comp.os.ms-windows.misc, comp.windows.x) where cross-posting occurs often between these three newsgroups. The second group, News-Related -3 consists of meassages from 3 related newsgroups (talk.politics.misc, talk.politics.guns and talk.politics.mideast). The third group, News-Different-3 contains 3 newsgroups of quite different topics (alt.atheism, rec.sport.baseball, and sci.space). Before use this dataset, we make a preprocessing, i.e., stemming, stop word removal.

**Table 1.** Selected News Groups

| Similar group | Related group | Different group |
| --- | --- | --- |
| comp.sys.ibm.pc.hardware | talk.politics.misc | alt.atheism |
| comp.sys.mac.hardware | talk.politics.guns | rec.sport.baseball |
| misc.forsale | talk.politics.mideast | sci.space |

### 5.2  Evaluation Criterion

We adopt two measurements to estimate cluster quality: normalized precision- recall measurement, and normalized mutual information (NMI) [22].

Let $CC_i$ be one of clusters, Let $CL_i$ be its corresponding class. Let $Corr_i$ be a set text from $CL_i$ that have been correctly assigned into $CC_i$ by out algorithm. Then the averaged precision and recall can be denoted as:

$$\Pr ecision = \frac{\sum_{i=1}^{k}|Corr_i|}{\sum_{i=1}^{k}|CC_i|}; \qquad \text{Re} call = \frac{\sum_{i=1}^{k}|Corr_i|}{\sum_{i=1}^{k}|CL_i|} \qquad (17)$$

The normalized mutual information measurement is defined as Eq.18, where $I(S;F)$ is the mutual information between cluster assignment $S$ and folder labels f, $H(S)$ is the entropy of S and $H(F)$ is the entropy of $F$, it measures the shared information between $S$ and $F$.

$$NMI = \frac{I(S;F)}{(H(S)+H(F))/2} \qquad (18)$$

These two measurements are correlated but show different aspects of clustering performance. *P-R* measurement calculates the relationship between individual and whole data; *NMI* measures the similarity between cluster partitions.

### 5.3  Results and Baseline Method

We make three contrast experiments with considered three baseline algorithms. We select the VSM algorithm, as well as probabilistic latent semantic analysis algorithm (PLSA) [12] and LDA algorithm --two popular generative models for unsupervised learning algorithm. Figure2 shows results of contrast experiments.

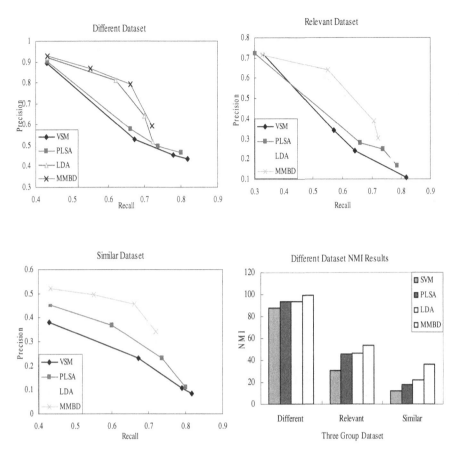

**Fig. 2.** Contrast Experiments Results on Different Datasets. Here, Figure (A)-(C) depicts the P-R Measurements Experiment Results. Figure (D) depicts the NMI Measurements Experiment Results.

Notice: in all datasets, the MMDB performs better than those baselines. But an interesting results for different datasets is that the MMDB and other baseline algorithms performance are consistent, i.e., on three different datasets of 20-Newsgroups, they performance are descending, the best is the different group dataset, the next is on related group dataset, the last is on similar group dataset. This maybe caused by the topic relevance.

In experiment, the PLSA algorithm is a classic probabilistic generative model, we can compile the PLSA code, according to Hoffman algorithm [12]. However, LDA algorithm is a novel probabilistic generative model, we write code and compile according to [23].

## 6   Conclusion and Future Work

In this paper, we proposed a bayesian mixture model, in which introduce a context variable, which has Dirichlet prior, in a bayesian framework to model text multiple topics and then clustering. It is a novel unsupervised text learning algorithm to cluster large-scale web data. In addition, parameters estimation we adopt Maximum Likelihood (ML) and EM algorithm to estimate the model parameters, and employed BIC principle to determine the number of clusters. Experimental results show that method we proposed distinctly outperformed baseline algorithms.

There are many future challenges, such as using active learning principles to optimize the summarization of a cluster, and building more sophisticated models to clustering large-scale data sets. So an important future research direction is to further study how to better the proposed mixture model in parameter estimation and judgments of cluster numbers, especially, how to incorporate the web user's opinion into our model, i.e., in objective mixture model we can acquire user's subjective interest or preference, which is very useful for facing future Web2.0. So in future work, we will study more sophisticated probabilistic models which can grasp user's latent sentiment. It is a novel and full of significant challenges.

## Acknowledgement

This work is supported by National Nature Science Foundation (No. 60573090).

## References

1. Liu, B., Lee, W.s., et al.: Partially supervised classification of text documents. In: Proc 19th Internet Conf. Machine Learning (ICML), pp. 387–394 (2002)
2. Liu, B., Dai, Y., et al.: Building text classifiers using positive and unlabeled examples. In: 3 rd IEEE Internet Conf. Data Mining(ICDM) (2003)
3. Bishop, C.M.: Pattern recognition and machine learning. Springer, Heidelberg (2006)
4. Sandler, M.: Hierarchical mixture models: a probabilistic analysis. In: Proc. of KDD 2007 (2007)
5. Sandler, M.: On the use of linear programming for unsupervised text classification. In: Proc. of KDD 2005 (2005)
6. Blei, D.M., Ng, A., Jordan, M.: Latent dirichlet allocation. Journal of Machine Learning Research 3, 993–1022 (2003)
7. Cheeseman, P., Stutz, J.: Bayesian classification (autoclass): theory and results[A]. In: Fayyad, U., Piatesky Shapiro, G., Smyth, P., Uthurusamy, R. (eds.) Advances in Knowledge Discovery and Data Mining[C], pp. 153–180. AAAI Press, Cambridge (1995)
8. Calvo, B., Larranaga, P., Lozano, J.A.: Learning Bayesian classifiers from positive and unlabeled examples. Pattern Recognition Letters (2007)
9. McLachlan, G., Basfor, K.: Mixture models, inference and applications to clustering. Marcel Dekker (1987)
10. Heller, K., Ghahramani, Z.: Bayesian hierarchical clustering. In: ICML (2005)
11. McCallum, A., Nigam, K.: A comparison of event models for naïve bayes text classification. In: AAAI, workshop on learning for text categorization (1998)
12. Hofmann, T.: Probabilistic Latent Semantic Analysis. In: The 22nd Annual ACM Conference on Research and Development in Information Retrieval, Berkeley, California, pp. 50–57. ACM Press, New York (1999)
13. Liu, X., Gong, Y., Xu, W., Zhu, S.: Document clustering with cluster refinement and model selection capabilities. In: Proc. of SIGIR 2002, pp. 191–198 (2002)

14. Li, H., Yamanishi, K.: Topic analysis using a finite mixture model. Information Processing and Management 39/4, 521–541 (2003)
15. Steyvers, M., Smyth, P., Rosen-Zvi, M., Griffiths, T.: Probabilistic author-topic models for information discovery. In: Proc. of KDD 2004, pp. 306–315 (2004)
16. Yu, Y., Xu, Q.F., Sun, P.F.: Bayesian clulstering based on fiinite mixture models of dirichlet distribution. [J] Mathematica Applicata 19(3), 600–605 (2006)
17. Mei, Q.Z., Ling, X., Wondra, M., et al.: Topic Sentiment Mixture: Modeling facets and opinions in weblogs. In: WWW 2007. ACM Press, Canada (2007)
18. Fraley, C., Raftery, A.E.: Model-based clustering discriminant analysis and density estimation[J]. Journal of the American Statistical Association 97, 611–631 (2002)
19. McLachlan, G.J., Peel, D.: Finite Mixture Models[M]. Wiley, New York (2000)
20. Basu, S., Bilenko, M., Mooney, R.J.: A probabilistic framework for semi-supervised clustering. In: Proc. of KDD 2004 (2004)
21. Huang, Y.F., Mitchell, T.M.: Text clustering with extended user feedback. In: Proc. of SIGIR 2006, ACM Press, Seattle, Washington USA (2006)
22. Dom, B.: An information-theoretic external cluster-validity measure. Technical Report RJ 10219, IBM (2001)
23. McCallum, A., Corrada-Emmanuel, A., Wang, X.: topic and role discovery in social networks. In: IJCAI -19, pp. 786–791 (2005)

# The Knowledge Discovery Research on User's Mobility of Communication Service Provider

Lingling Zhang[1,2], Jun Li[1,2,*], Guangli Nie[1], and Haiming Fu[1]

[1] School of Management, Graduate University of Chinese Academy of Sciences,
Beijing, 100080, China
[2] Chinese Academy of Sciences Research Center on Fictitious Economy and Data Sciences,
CAS, Beijing, 100080, China

**Abstract.** This paper built two models based on the existing data of a mobile communication service provider. In the modern life, mobile phone has already become a necessity for us to take at any time. The position change of the mobile phone holder represents the cellular phone user's move. A lot of promotion and other business activities are related to user's position, and many business activities need to know the customers' long-term move regulation. Therefore, the research of the cellular holder's move principle is worthy now. The cellular phone customer move analysis of the paper is an analytical system based on the customer position. We conducted analysis for single time of the cellular phone customer ordering of position. We carried on analysis of several time positions and the relation of these positions. The mobile regulation of customer is analyzed by analyzing the relation of these positions. Different business activity can carry on promotion to the customer based on different regulation. According to the detection of the customer mobile regulation, the enterprise can discover new business application, and make more effectively business activities.

**Keywords:** Knowledge discovery, communication service provider, move of position.

## 1 Introduction

The cellular phone holders are mobile individuals, and move all the time everyday. Each mobile individual represents a kind of behavior pattern and a kind of consumption pattern. For example, the consumption pattern of the person who goes to work by car is different from the person who goes to work on foot (Saharon Rosset et al., 2004). The consumption view of the person who goes home after work everyday is different from the person who has a rich night life after work finished. The person who has different lifestyle would need different mobile service. Therefore, the mobile characters of customers have already become a typical factor to distinguish the people with different consumption view, consuming way and consuming ability.

---

* Corresponding author.

Y. Ishikawa et al. (Eds.): APWeb 2008 Workshops, LNCS 4977, pp. 182–191, 2008.

In brief, mobility is a way to differentiate the mobile customer, including everyday route of the mobile holder, the move scope of the mobile holder, mobility speed and so on. We can get the activity rule and the living habit and learn the users' living location and working position through the mobility analysis of the customers' daily move data. Summarizing these classifications of the users, different marketing strategies can be conducted to different users. Based on the analysis about users' behavior characteristic, the communication service providers can find the potential market and develop new profitable business. In the past, because of lacking of the effective data analysis and data support, when the providers designed company marketing strategy, they often weren't clear who their real marker targets were. Due to the lack of in-depth analysis and insight, they used to design service suites from themselves' perspective instead of the customers'. Moreover, behavior analysis on the cell phone users can also find many new business needs, thereby enhancing customer satisfaction (Jaakko Hollmen,Volker Tresp,2005).

The communication service providers of China (taking China Mobile Communication Corporation for example) analyze the customers mainly on BASS (Business Analyses Support System). BASS is built based on computer network and data warehouse technology. The analysis area of this system covers almost all the business of communication service providers (T. PAGE, 2005). But due to: 1) because different departments have different data sources, the data of mobile users is rather complex. Generally there are four main sources, which are separately named BOSS systems, customer service systems, MISC system, and network management system. 2) The volume of mobile users' data of is huge. 3) There are many factors that impact the mobile phone users, and they are interrelated and extremely complex.

Therefore, the generally simple statistical analysis or OLAP analysis couldn't find the most essential reasons causing the problems. Using the more appropriate analytical tools and algorithm to discover knowledge from a large number of data to manage the customers has become critical issue for the mobile service provider.

As Information Technology (IT) progresses rapidly, many enterprises have accumulated large amounts of transaction data; how to extract available implicit knowledge to aid decision making from this data has became a new and challenging task. (KDD) The term first appeared in 1989 at the 11th International Joint Conference on Artificial Intelligence(Usama Fayyad,Paul Stolorz,1997). This new business information processing technology, according to established commercial business goals, can take the large number of commercial data exploration and analysis and reveal hidden, unknown or known to test the law, and further to discover knowledge from the database. The knowledge mined always is unknown before and is tacit knowledge. When extracting knowledge from data, there are various problems has to deal with including characterization, comparison, association, classification, prediction and clustering.

The rest of the paper is organized as follows. The second section reviewed the related research on the analysis about mobile users. Knowledge discovery research model about user mobility is built in the third section. We discussed commercial value and application of the mobile move analysis in section 4. Finally, the conclusion is given in section 5.

## 2   Review of Research Related to Mobile Users' Mobility

At present, study related to the mobility of mobile users at home and abroad stays in two phases. Stage one is cell phone users' location identification; the second stage is mobile phone users' analysis. We describe the technical features of the two-stage in the following section. (S.C.Hui,G.Jha, 2000; Saharon Rosset ,et al., 2004).

The first stage is about cell phone users' location identification. This stage is the initial stage of mobile users' analysis, which is to identify the location of cell phone users. The technology of this phase is the position recognition, which is passive position identification. Currently in China, China Mobile and China Unicom will launch its own mobile network based on the positioning of mobile phone services. The GpsOne service provided by China Unicom is with higher precision, which is up to 5 meters. But no matter how the service positions accurately, the two companies only provide the position service to a specific cell phone user group instead of all the mobile phone users. Location identification refers to that the mobile service providers inform the user where they are when they require service.

The second stage is the analysis about Mobile phone users' move regulation. The analysis in this stage is a deeper analysis about the cell phone users' location, which study the mobile habits of the users so as to reach a more valuable conclusion. Typical example is the UK Vodafone (UK mobile co Vodafone) company. Vodafone company analyzed the call records of the customers to study the communication habit of the customers. They found the people who are inclined to call within a near area, and other people who are inclined to dial long-distance call. (Ali Kamrani,Wang Rong,Ricardo Gonzalez,2001).

There are a lot of shortcomings to improve in the existing analysis no matter about mobile phone users' move rule or the application. For example:

1). How to avoid adding new location identification system cost during analysis;
2). How to reduce the mobile communication system resource consumption and the impact on system load;
3). How to do continuity location identification to large-scale mobile phone users;
4). How to avoid the uneven distribution of the positioning time;
5). How to change mobile users' analysis results into effective business applications.

Due to the above reasons, we need to find a new method which identify the location of cell phone users efficiently, and accordingly generating a mobile user model which can be applied to a commercial service effectively.

## 3   Knowledge Discovery Research of User Mobility Model

### 3.1  The Objective of Mobile Users' Analysis and Data Preprocess

The objective of mobile users' analysis is to identify users' activity character or activities law, which can deduce users' living habits, consumption habits and consumer attitudes. The model generated based on these theories would offer the enterprise related to communication very useful decision making reference. The business value of this analysis is broad, such as promoting new business, cross-selling, marketing, advertising, psychological analysis of user groups, and prediction about the users' need. The goal of this paper embodies as follows.

1). Identify the users' regional activities;
2). Identify the user's activities law and character (in reference to the time of the activities of regional and mobile speed);
3). Identify the location of user' work and apartment;
4). Identify the users' long-term move routes;
5). Identify the users' long-term move character.

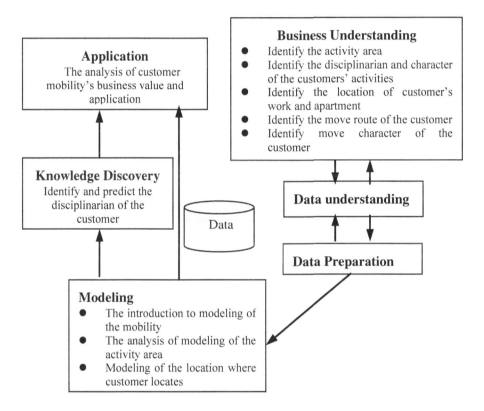

**Fig. 1.** The process of knowledge discovering

**Data collection and transform:** The original data collected at the beginning is recorded in binary system. The text formatted data can be generated through Agilent which is a tool. After cleaning and transforming the record, the information related to the location is in order. The original data can be dropped.

**Data sampling:** The sample consists of 24881 mobile service customers sampled from all of the customers of a city in China.

We set the span of the time as following.

Morning : 06 : 00−09 : 00  Working : 09 : 00−17 : 00
Evening : 17 : 00−22 : 00  Night/ : 22 : 00−06 : 00

## 3.2  User Mobility Modeling

As the development of artificial intelligence and database technologies, knowledge discovery is an important cross-discipline way used in the past 10 years. The algorithm will find the implicit knowledge hidden in the data and provide support for the expert decision.

Cluster analysis refers to a different sample classification, and it is a commonly used statistical methods. When the analyzed data lacks of described information, or it is unable to form any classification model, cluster analysis is quite useful then. Cluster analysis is based on a similar level measurement method. The data in the same cluster should be most similar and the difference between clusters should be most large. Clustering methods include statistical methods, machine learning methods.

Statistical method of cluster analysis is a mean to achieve clustering, mainly based on the geometric distance among the samples. The machine learning clustering concept is based on the description. Description concept is targeted to certain types of content description and a general object of this feature. Concept description is divided into characteristic description and differential description. The former describes the common feature of certain objects and the latter describes different types of distinction between the object.

There are a lot of attributes in the actual data about the mobile users, including price attributes, behavioral attributes and the natural attributes etc., and each of the attributes comprise of several parameters. Ambiguity and complexity of the real world make it difficult for us to decide that how many categories that the user should set into. Then clustering algorithm can be used to allow machines to help us detect the categories of the users. We build two models in this section.

### 3.2.1  Model A: Modeling Analysis about Users' Activity Area

This kind of model mainly analyzes the users' mobile distance in various times to mark the regional users, and analyzes users' mobile features. After preprocessed, derived data generated from the original data would reflected the largest mobile distance in the morning, between morning and working hours, work hours, between work and evening, in the evening, between the evening and deep at night, deep at night and whole day.

In this paper, we used the Cluster Analysis K of SPSS to analyze the data- Means clustering algorithm analysis. The main parameters are set as follows:

Variables: Morning; Inter1; Working; Inter2; Evening; Inter3; Night; Day
Number of Clusters: 12
Maximum Iterations: 30
Convergence Criterion: 0.02
Clustering results are shown in Table 1
Final Cluster Centers for User Active-Distance Clustering:

**Table 1.** Users' mobility attribute clustering center

| Time Span | Cluster | | | | | | | | | | | |
|---|---|---|---|---|---|---|---|---|---|---|---|---|
| | 1 | 2 | 3 | 4 | 5 | 6 | 7 | 8 | 9 | 10 | 11 | 12 |
| Morning | 3.1 | 3.5 | 1.6 | 13.3 | 1.0 | 14.2 | 1.8 | 15.5 | 14.6 | 4.2 | 3.5 | 15.1 |
| inter1 | 2.4 | 1.6 | 2.5 | 4.0 | 0.6 | 6.5 | 0.9 | 10.2 | 6.0 | 2.1 | 2.6 | 5.4 |
| Working | 16.2 | 5.7 | 14.7 | 3.5 | 1.9 | 16.2 | 3.2 | 16.7 | 16.5 | 12.0 | 15.6 | 17.2 |
| inter2 | 1.7 | 1.7 | 1.3 | 1.5 | 0.6 | 2.8 | 3.5 | 5.4 | 2.3 | 4.0 | 14.6 | 6.8 |
| Evening | 15.0 | 3.8 | 1.8 | 2.3 | 1.3 | 2.2 | 12.6 | 15.2 | 2.4 | 14.4 | 10.5 | 16.2 |
| inter3 | 1.1 | 3.1 | 0.8 | 1.1 | 0.5 | 8.1 | 1.1 | 3.9 | 2.1 | 14.0 | 1.5 | 8.3 |
| Night | 2.3 | 11.1 | 0.9 | 1.2 | 0.6 | 14.9 | 1.0 | 2.0 | 1.4 | 3.9 | 1.8 | 15.5 |

## Explanation to the Model

We present the cluster results with the following figures in a more direct way. The X axis of below figures presents time which is from morning to night, and the Y axis is the distance the mobile phone moved, and the unit of the Y axis is KM.

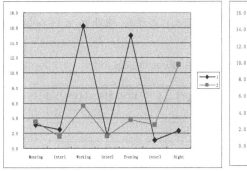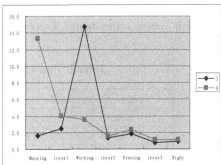

**Fig. 2.** Mobile users' cluster model

Cluster 1: From the the blue line of the left figure, we can get the user has a large activity scope range in 14-16Km during the working time and the evening.

Cluster 2: From the pink line of the left figure, we can get the user has a large activity scope during the night, and the objective has smaller activity scope at other time.

Cluster 3: From the blue line of the right figure, we can get the user has a large activity scope range above 14Km during working time and has smaller activity scope below 2Km during other time.

Cluster 4: The pink line of the right figure, we can get the user has a large activity scope during the morning, and has smaller activity scope at other time.

For the sake of space, we won't illustrate other eight statuses in detailed.

### 3.2.2  Model B: Mobile Users Line Modeling Analysis

Another way of analyzing the mobility of users is to identity the users with special moving feature. For example, by analyzing the route of the users, who usually linger at the entertainment street and who usually drive on the Shanghai-Hangzhou expressway. This modeling process is different from above clustering modeling analysis, because it is a classification model.

Training data is the first factor of building classification model. Take identifying the characteristics of users who drive on Shanghai-Hang Zhou expressway as an example, we first have to know about features of the users driving on the road. The original data can record where the mobile phone user is in different time, and the combination of time and location can express the attributes of the travel route.

In light of the actual situation, we may consider several key elements of attributes in the model:

1) Position sequence: On the path of the relative location, the point the record can only be sequenced in a fixed order.
2) Time relation: The time needed when the users past some points is the function of the speed, so the time relations were relatively fixed.

Based on the above analysis, we expressed the character of certain with numerical parameter.

This is a relevant time series modeling problem. First we collect one group of the data, get the path at various points in time sequence as shown in Table2. We will establish as time series model. There are a, b, c, d, e, f, g, h, I, j, k, l, m, n 14 points in

**Table 2.** Time series of a Certain Route Model

| Time | Cell | Spot | Time_Spend (sec) |
|---|---|---|---|
| 11:28:30 | 22556-11722 | a | 0 |
| 11:30:43 | 22556-11582 | b | 132 |
| 11:32:01 | 22556-11111 | c | 210 |
| 11:32:44 | 22556-11911 | d | 253 |
| 11:33:15 | 22556-31261 | e | 285 |
| 11:34:02 | 22556-21921 | f | 332 |
| 11:34:52 | 22556-11921 | g | 382 |
| 11:35:07 | 22556-31921 | h | 397 |
| 11:35:26 | 22307-21101 | i | 416 |
| 11:36:24 | 22307-11101 | j | 474 |
| 11:38:42 | 22307-11221 | k | 612 |
| 11:39:55 | 22307-31171 | l | 685 |
| 11:40:55 | 22307-21171 | m | 745 |
| 11:41:30 | 22307-11171 | n | 780 |

**Table 3.** Route X time series

| Time | Cell | Spot | Time_Spent(sec) |
|------|------|------|-----------------|
| 12:32:23 | 22556-11722 | a | 0 |
| 12:34:07 | 22556-11582 | b | 104 |
| 12:36:12 | 22556-11111 | c | 229 |
| 12:36:52 | 22450-11931 | | |
| 12:37:00 | 22556-21921 | f | 277 |
| 12:38:17 | 22556-11921 | g | 354 |
| 12:38:38 | 22556-31921 | h | 375 |
| 12:39:00 | 22307-21101 | i | 396 |
| 12:40:03 | 22307-11101 | j | 459 |
| 12:42:11 | 22307-11221 | k | 588 |
| 12:43:42 | 22307-31171 | l | 679 |
| 12:44:38 | 22307-21171 | m | 735 |
| 12:45:18 | 22307-11171 | n | 775 |

the series. Deleted some repeated points. Another time series shown in Table3 is a path of non-standard time series. Some points are missing in the trail, but the timing sequence is correct. There are a, b, c, f, g, h, I, j, k, l, m, n 12 points in Rout X, where d, e is missing. We will find a way to make it be identified correctly.

Time series conversion: Take the first point path as a reference point, other points occurred in the following time is comparison point, and calculates time difference between these points and the first referenced time. Take Time_Spent list as Y axis, and Spot list as X axis, Figure 3 is the figure showing the route of the model. Similarly, the time series of Table 3 can be converted into X-Y coordinates figure, as Figure 3 showed. The missing points are presented as blank in the chart.

We identify the two graphics are the same path through this recognition algorithm.

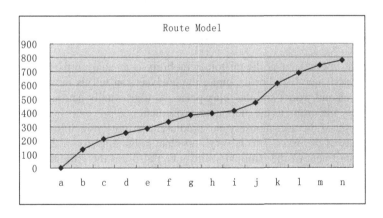

**Fig. 3.** Route Model route changed chart

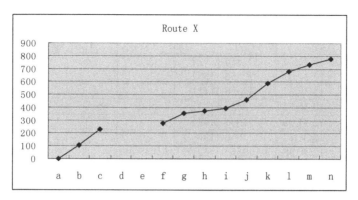

**Fig. 4.** Route X route changed chart

# 4 The Business Value and Application of Users' Mobility

### 4.1 The Application of the Mobility Service Provider

The analysis of the mobility of users could be very helpful to the mobility service provider. The mobility service provider can make marketing strategy as to the character of the mobility service users. They also can modify their crossing selling strategy.

Making marketing strategy: the mobile service provider can recognize the place where they can attract more customers. Because of the limitation of the purchasing channel, the place where is suitable to attract new customers may not be the best place to set up a service center. The mobility service provider can estimate the place where is profitable and adjust the service center as to the estimation result.

Crossing selling strategy: the mobile service provider can focus on some selling based on the mobility attribute of the customer. For example, the mobile service provider can fortify the promote to the people who spend a lot of time on the way to work or home because they may choose the service of the mobile TV service.

### 4.2 The Application of the Mobile Advertisement and National Security

An important application of the customer mobility is to analyze the related attribute about the users' mobility. For example, the analysis can tell us the place where the customer lives and works and the relax place the user usually patronizes. This information is very important to the mobile advertisement.

The analysis of the user mobility can predict the activity principle of the users and track the route of the users. This is very important to the ministry of the national security. The department of the national security can also tail the criminal suspicion and even detect the potential crime as to the activity route.

# 5 Conclusions

In sum, we can develop several models through the analysis of the user mobility. We can get the route, the mobility disciplinarian of the users and the place where the user linger. This information can support the commercial decision of the mobile service provider.

Form the perspective of the commerce, the user models and the related analysis can provide technology support. Besides the promotion of new service, the cross selling, the location of the users, the model can be helpful to the creation of the new business model.

The date source of the analysis of the user mobility is valuable. The potential value of the data has not been completely mined. More models can be developed based on the mobility data; this is the future work of this paper.

## Acknowledgements

This research has been partially supported by grants from National Natural Science Foundation of China (No. 70501030 and Innovation Group 70621001), and Beijing Natural Science Foundation of (No.9073020).

## References

1. Page, T.: A Classification of Data Mining Techniques, First Quarter 2005 SME Technical Papers (2005)
2. Kiang, M.Y., Fisher, D.M., Fisher, S.A., Chi, R.T.: Understand Corporate Rationales for Engaging in Reverse Stock Splits - A Data Mining Application. In: The 38th Annual Hawaii International Conference on System Sciences (HICSS 2005) (2005)
3. Yaik, O.B., Yong, C.H., Haron, F.: Time series prediction using adaptive association rules. In: Proceedings of the International Symposium on Information Theory, ISIT 2004 (2005)
4. Kamrani, A., Rong, W., Gonzalez, R.: A genetic algorithm methodology for data mining and intelligent knowledge acquisition. Computers and Industrial Engineering 40(4), 361–377 (2001)
5. Fayyad, U., Stolorz, P.: Data mining and KDD: Promise and challenges. Future Generation Computer Systems 13(2-3), 99–115 (1997)
6. Hollmen, J., Tresp, V.: Call-based Fraud Detection in Mobile Communication Networks using a Hierarchical Regime-Switching Model
7. Rosset, S., Murad, U., Neumann, E., Idan, Y., Pinkas, G.: Discovery of Fraud Rules for Telecommunications-Challenges and Solutions
8. Hui, S.C., Jha, G.: Data mining for customer service support. Information & Management (38), 1–13 (2000)

# Protecting Information Sharing in Distributed Collaborative Environment*

Min Li and Hua Wang

Department of Mathematics & Computing
University of Southern Queensland, Australia
{limin,wang}@usq.edu.au

**Abstract.** Information sharing on distributed collaboration usually occurs in broad, highly dynamic network-based environments, and formally accessing the resources in a secure manner poses a difficult and vital challenge. Our research is to develop a systematic methodology for information sharing in distributed collaborative environments. It will ensure sensitive information and information assurance requirements, and incorporate new security constrains and policies raised by emerging technologies. We will create a new rule-based framework to identify and address issues of sharing in collaborative environments; and to specify and enforce security rules to support identified issues while minimizing the risks of information sharing through the framework.

## 1 Aims and Background

We aim to develop a policy-based framework for information sharing in distributed collaborative environments with role-based delegation and revocation. The motivation of role-based delegation and revocation are that users themselves may delegate role authorities to others to process some authorized functions and later remove the authorities. Role-based delegation and revocation models will be developed with comparisons to established technical analysis, laboratory experiments, support hierarchical roles and multistep delegation. An innovation policy-based language for specifying and enforcing rules on the models is proposed as the fundamental technique within this framework. The models will be implemented to demonstrate the feasibility of the framework and secure protocols for managing delegations and revocations.

Delegation is the process whereby an active entity in a distributed environment grants access resource permissions to another entity. In today's highly dynamic distributed systems, a user often needs to act on another user's behalf with part of the user's rights. To solve such delegation requirements, ad-hoc mechanisms are used in most systems by compromising existing disorganized policies or additional components to their applications [26, 16, 18]. The basic idea of delegation is to enable someone to do a job, for example, a secretary.

---

* The research is support by an ARC Discovery Grant DP0663414.

Y. Ishikawa et al. (Eds.): APWeb 2008 Workshops, LNCS 4977, pp. 192–200, 2008.

Effective delegation not only makes management systems ultimately more satisfactory, but also frees the delegating users to focus on other important issues. In access control management systems, the delegation arises when users need to act on another user's behalf in accessing resources. The delegation might be for a short time, for example, sharing resources temporarily with others during one week holiday. Otherwise users may perceive security as an obstacle of the resources sharing. With delegation, the delegated user has the privileges to access information without referring back to the delegating user.

Delegation is recognized as vital in a secure distributed computing environment [1, 3, 10]. The most common delegation types include user-to-machine, user-to-user, and machine-to-machine delegation. They all have the same consequence, namely the propagation of access permission. Propagation of access rights in decentralized collaborative systems presents challenges for traditional access mechanisms because authorization decisions are made based on the identity of the resource requester. Unfortunately, access control based on identity may be ineffective when the requester is unknown to the resource owner. Recently some distributed access control mechanisms have been proposed: Lampson *et al.* [12] present an example on how a person can delegate its authority to others; Blaze *et al.* [5,6],introduced trust management for decentralized authorization; Abadi *et al.* [1] showed an application of express delegation with access control calculus; and Aura [2] described a delegation mechanism to support access management in a distributed computing environment.

The National Institute of Standards and Technology developed role-based access control (*RBAC*) prototype [7] and published a formal model [9]. *RBAC* enables managing and enforcing security in large-scale and enterprise-wide systems. Many enhancements of *RBAC* models have been developed in the past decade. In *RBAC* models, permissions are associated with roles, users are assigned to appropriate roles, and users acquire permissions through roles. Users can be easily reassigned from one role to another. Roles can be granted new permissions and permissions can be easily revoked from roles as needed. Therefore, *RBAC* provides a means for empowering individual users through role-based delegation in distributed collaboration environments.

The importance of delegation has been recognized for a long time, but the concept has not been supported in *RBAC* models [8, 19]. A security officer has to assign a role to the delegated user if the role is required to be delegated to the user. Such a model significantly increases the management efforts in a decentralized collaboration environments because of the dynamic of delegations and the continuous involvement from security officers. We will provide a bridge of the gap between delegation techniques and *RBAC* models.

## 2   Significance and Innovation

Delegation is an important feature in many collaboration applications. For example, the Immigration Department is developing partnerships between immigration agencies and people in local areas to address possible problems.

Immigration officers are able to prevent illegal stay and crime if they efficiently collaborate with the people. The problem-oriented immigrating system (*POIS*) is proposed to improve the service as a part of the Immigration Department's ongoing community efforts including identifying potential problems and resolving them before they become significant. With efficient delegation, officers respond quickly to urgent messages and increase the time spent confronting problems.

In *POIS*, officers might be involved in many concurrent activities such as conducting initial investigations, analyzing and confronting crimes, preparing immigration reports, and assessing projects. In order to achieve this, users may have one or more roles such as lead officer, participant officer, or reporter. In this example, Tony, a director, needs to coordinate analyzing and confronting crimes and assessing projects. Collaboration is necessary for information sharing with members from these two projects. To collaborate closely and make two projects more successful, Tony would like to delegate certain responsibilities to Christine and her staff. The prerequisite conditions are to secure these processes and to monitor the progress of the delegation. Furthermore, Christine may need to delegate the delegated role to her staff as necessary or to delegate a role to all members of another role at the same time. Without delegation skill, security officers have to do excessive work since the involvement of every single collaborative activity. We can find the major requirements of role-based delegation in this example:

1. Group-based delegation means that a delegating user may need to delegate a role to all members of another role at the same time. We introduce a new ability-based delegation model in our recent work [13].
2. Multistep delegation occurs when a delegation can be further delegated. Single-step delegation means that the delegated role cannot be further delegated.
3. Revocation schemes are important characters in collaboration. They take away the delegated permissions. There are different revoking schemes, among them are strong and weak revocations, local and global revocation. We discuss these different revocation with according algorithms in our recent paper [14].
4. Constraints are an important factor in *RBAC* for laying out higher-level organizational policies. It defines whether or not the delegation or revocation process is valid.
5. Partial delegation means only subsets of the permissions are delegated while total delegation means all permissions are delegated. Partial delegation is an important feature because it allows users only to delegate required permissions. The well-known least privilege security principle can be implemented through partial delegation.

Although the concept of delegation is not new in authorizations [2, 3, 5, 10, 16, 17, 21, 25], role-based delegation received attention only recently [15, 27, 28]. Aura [2] introduced key-oriented discretionary access control systems that are based on delegation of access rights with public-key certificates. A certificate has the meaning:

$S_K$ (During the validity period, if I have the rights $R$, I give them to someone).

$S_K$ denotes a signed message that includes both the signature and the original message. The key that signed the certificate ($K$) is the issuer and the rights $R$ given by the certificate are the authorization. With the certificate, the issuer delegates the rights $R$ to someone. The systems emphasized decentralization of authority and operations but their approach is a form of discretionary access control. Hence, they can neither express mandatory policies like Bell-LaPadula model [4], nor possible to verify that someone does not have a certificate. Furthermore, some important policies such as separation of duty policies cannot be expressed with only certificates. They need some additional mechanism to maintain the previously granted rights and the histories must be updated in real time when new certificates are issued. Delegation is also applied in decentralized trust management [6,15,16]. Blaze *et al.* [6] identified the trust management problem as a distinct and important component of security in network services and Li *et al.* [15,16] made a logic-based knowledge representation for authorization with tractable trust-management in large-scale, open, distributed systems. Delegation was used to address the trust management problem including formulating security policies and security credentials, determining whether particular sets of credentials satisfy the relevant policies, and deferring trust to third parties. Other researchers have investigated machine to machine and human to machine delegations [1,10,25]. For example, Wang *et al.* [25] proposed a secure, scalable anonymity payment protocol for Internet purchases through an agent which provided a higher anonymous certificate and improved the security of consumers. The agent certified re-encrypted data after verifying the validity of the content from consumers. The agent is a human to machine delegation which can provide new certificates. However, many important role-based concepts, for example, role hierarchies, constraints, revocation were not mentioned.

Wang *et al.* [21] discussed the mobility of user-role relationship in *RBAC* management and provided new authorization allocation algorithms for *RBAC* along with the mobility that are based on relational algebra operations. They are the authorization granting algorithm, weak and strong revocation algorithms. The paper does not use role delegation but instead defines the role mobility, whereby a user with an mobile role may further grant other roles but she/he cannot accept other roles if she/he has an immobile role. The mobility could be viewed as a special case of multistep delegation in their work. But some important delegation features such as partial delegation and delegation revocation have not been considered. Barka and Sandhu [3] proposed a simple model for role-based delegation called *RBDM0* within *RBAC0*, the simplest form of *RBAC96* [19]. *RBDM0* is a simple delegation model supporting only flat roles and single step delegation. However, they neither gave the definition of role-based delegation relation, which is a critical notion to the delegation model nor discussed the relationships among original user and delegated user. Some important features such as role hierarchies and revocations were not supported in *RBDM0*.

Some researchers have worked on the semantics of authorization, delegation, and revocation. Wang *et al.* [20] described a secure and flexible protocol and its role based access control for M-services. The protocol is based on a Credential Center, a Trusted Center and a ticket based mechanism for service access. It supports efficient authentication of users and service providers over different domains and provides a trusted model for participants. The concepts, protocols, and algorithms for access control in distributed systems from a logical perspective have been studied. However, there is no multistep delegation control mechanism since every delegation can be freely delegated. Hagstrom *et al.* [11] studied various problems of revoking in an ownership-based framework, but their attempt was still not sufficient to model all the revocations required in role-based delegation, for example, grant-independent and duration-restricted revocations. Zhang *et al.* [27, 28] proposed a rule-based framework for role-based delegation including *RDM2000* model. *RDM2000* model is based on *RBDM0* model with some limitations that we mentioned before. Furthermore, as a delegation model, it does not support group-based delegation. *RDM2000* does not consider administrative role delegation but the deletion of regular roles. The model does neither analyse how do original role assignment changes impact delegations nor implement with XML-based language.

We will focus exclusively on how to specify and enforce policies for authorizing role-based delegation and revocation using a rule-based language. We will continue our previous work and propose delegation frameworks including revocation models, group-based, multistep, and partial delegations. With the revocation models, we will not only consider the deletion of regular roles but also administrative role delegation. Additionally, in order to provide sufficient functions with the framework, we will analyze how do original role assignment changes impact delegations and implement with XML-based language. This kind of language for role-based delegation has not been studied.

## 3   Approach

### Task 1: A Role-Based Delegation Framework

This task will develop a delegation framework called *RBDF*. This framework supports role hierarchy and multistep delegation and revocation by introducing the delegation relation, delegation authorization, role-based revocation and revocation authorization.

Two relations are included in role-based access control: user-role assignment (*URA*) and permission-role assignment (*PRA*). *URA* is a many-to-many relation between users and roles and *PRA* is a many-to-many relation between permissions and roles. Users are enabled to use the permissions of roles assigned to them. *RBAC* management systems have many advantages with its flexibility of assigning permissions to roles and users to roles [24]. There are two types of roles associated with user: *Original roles* and *Delegated roles*. The former is a role assigned to the user whilst the latter one is a role delegated to the user.

The same role can be an original role of one user and a delegated role of another user. Role hierarchy is a basic relationship between roles that specifies which role may inherit all of the permissions of another role. The relationship of

*Senior-Junior* shows hierarchies between roles. Senior roles inherit permissions from junior roles. Role hierarchies provide a powerful and convenient means to satisfy the least privilege security principle since only required permissions are assigned to a role. Because of role hierarchies, a role may be an original role and a delegated role of the same user. The original user-role assignment ($OUA$) is a many-to-many user-role assignment relation between users and original roles. The delegated user-role assignment ($DUA$) is a many-to-many user assignment relation between users and delegated roles.

**Role-Based Delegation.** Relational database systems will be designed. Database systems have been applied in our previous work to solve consistency problems in user-role assignment and permission-role assignment [22,23]. A set of relations such as *roles, users, permissions, user-role, role-permission* has been developed [23,24] for the formal approaches that are based on relational structure and relational algebra operation in database system. There are three major elements in a delegation relation: original user-role assignments ($OUA$), delegated user-role assignment ($DUA$), and constraints. Constraints are very important in role-based model [21]. Delegation may associate with zero or more constraints. The delegation relation supports partial delegation in a role hierarchies: a user who is authorized to delegate a role $r$ can also delegate a role that is junior to $r$.

As we mentioned before, there are various delegations in real-time application: single-step, multistep, group-based, and partial delegations. In single-step delegation the delegated role cannot further delegate. We also can define a maximum number of steps in multistep delegation. The maximum delegation number imposes restriction on the delegation. Single-step delegation is a special case of multistep delegation with maximum delegation number equal to one. We will develop delegation models to support these different delegations.

**Delegation Authorization.** The delegation authorization goal imposes restrictions on which role can be delegated to whom. We partially adopt the notion of prerequisite condition from Wang *et al.* [23] to introduce delegation authorization in the rule-based delegation framework ($RBDF$).

We will develop database systems for $RBDF$ in this task to support group-based, multistep, partial delegations and revocations and analyze what delegation impact will happen if an original role assignment is changed.

## Task 2: The Rule-Based Policy Specification Language

The motivation behind policy-based language are: 1) delegation relations defined in role-based delegation model lead naturally to declarative rules; 2) an individual organization may need local policies to further control delegation and revocation. A policy-based system allows individual organizations to easily incorporate such local policies.

We will show how our construction is used to express delegation and revocation policies.

**The Language.** The rule-based specification language specifies and enforces authorization of delegation and revocation based on the new delegation model.

It is entirely declarative so it is easier for security administrators to define policies. The proposed language will be a rule-based language with a clausal logic. A clause, also known as a rule, takes the form: $H \leftarrow B$. where $H$ stands for rule head and $B$ stands for rule body. $B$ is a prerequisite condition of a successful $H$. If the condition defined in the rule body is true, then it will trigger authorizations. An advantage is that the rule body can include the condition of an authorization policy and the rule head can include the authorization. This provides the mechanism for authorization specification and enforcement.

**Rules for Enforcing Policies.** Basic authorization will specify the policies and facts in the delegation framework. Addition to the basic authorization policies, further derivations are needed for authorization and their enforcement. A derivation rule body describes a semantic logic that consists of basic authorization, conditions and functions. The result can be either authorized or denied.

The language developed in Task 2 will be used in the database systems (Task 1) to process delegation and revocation authorizations.

## 4    Current Progress

1. We develop a flexible ability-based delegation model (ABDM), in which a user can delegate a collection of permissions, named an ability, to another user or all members of a group; we also analyze delegation granting and revocation authorization algorithms in this model [13]. (Part of Task 1)
2. we discuss granting and revocation models related to mobile and immobile memberships between permissions and roles and provide proposed authorization granting algorithm to check conflicts and help allocate the permissions without compromising the security [14]. (Part of Task 1)
3. We specify constraints of Usage Control Model (UCON) with object constraints language (OCL). The specification not only provides a tool to precisely describe constraints for system designers and administrators, but also provides the precise meaning of the new features of UCON, such as the mutability of attributes and the continuity of usage control decisions. This work is under preparation for submitting. (Part of Task 2)

## References

1. Abadi, M., Burrows, M., Lampson, B., Plotkin, G.: A calculus for access control in distributed systems. ACM Trans. Program. Lang. Syst. 15(4), 706–734 (1993)
2. Aura, T.: Distributed access-rights management with delegation certificates. In: Vitec, J., Jensen, C. (eds.) Security Internet programming, pp. 211–235. Springer, Berlin (1999)
3. Barka, E., Sandhu, R.: A role-based delegation model and some extensions. In: Proceedings of 16th Annual Computer Security Application Conference, Sheraton New Orleans, December 2000a, pp. 168–177 (2000)
4. Bell, D.E., La Padula, L.J.: Secure Computer System: Unified Exposition and Multics Interpretation, Technical report ESD-TR-75-306, The Mitre Corporation, Bedford MA, USA (1976)

5. Blaze, M., Feigenbaum, J., Lacy, J.: Decentralized trust management. In: IEEE Symposium on Security and Privacy, Oakland, CA, pp. 164–173 (1996)
6. Blaze, M., Feigenbaum, J., Ioannidis, J., Keromytis, A.: The role of trust management in distributed system security. In: Vitec, J., Jensen, C. (eds.) Security Internet Programming, pp. 185–210. Springer, Berlin (1999)
7. Feinstein, H.L.: Final report: NIST small business innovative research (SBIR) grant: role based access control: phase 1. Technical report. SETA Corporation (1995)
8. Ferraiolo, D., Cugini, J., Kuhn, D.R.: Role-based access control (RBAC): features and Motivations. In: Proceedings of 11th Annual Computer Security Application Conference, New Orleans, LA, December, pp. 241–241 (1995)
9. Ferraiolo, D.F., Kuhn, D.R.: Role based access control. In: The proceedings of the 15th National Computer Security Conference, pp. 554–563 (1992)
10. Gladney, H.: Access control for large collections. ACM Transactions on Information Systems 15(2), 154–194 (1997)
11. Hagstrom, A., Jajodia, S., Presicce, F., Wijesekera, D.: Revocations-a classification. In: Proceedings of 14th IEEE Computer Security Foundations Workshop, Nova Scotia, Canada, June, pp. 44–58 (2001)
12. Lampson, B.W., Abadi, M., Burrows, M.L., Wobber, E.: Authentication in distributed systems: theory and practice. ACM Transactions on Computer Systems 10(4), 265–310 (1992)
13. Li, M., Wang, H., Plank, A.: ABDM: An Extended Flexible Delegation Model in RBAC. In: IEEE-CIT 2008 (accepted, 2008)
14. Li, M., Wang, H., Plank, A.: Algorithms for advanced permission-role relationship in RBAC. In: ACISP 2008 (submitted, 2008)
15. Li, N., Feigenbaum, J., Grosof, B.N.: A logic-based knowledge representation for authorization with delegation (extended abstract). In: Proceeding 12th intl. IEEE Computer Security Foundations Workshop, Italy, pp. 162–174 (1999)
16. Li, N., Grosof, B.N.: A practically implementation and tractable delegation logic. In: IEEE Symposium on Security and Privacy, pp. 27–42 (May 2000)
17. Liebrand, M., Ellis, H., Phillips, C., Ting, T.C.: Role delegation for a distributed, unified RBAC/MAC. In: Proceedings of Sixteenth Annual IFIP WG 11.3 Working Conference on Data and Application Security King's College, University of Cambridge, UK July, pp. 87–96 (2002)
18. Mcnamara, C.: Basics of delegating. (1997), http://www.mapnp.org/library/guiding/delegate/basics.htm
19. Sandhu, R., Coyne, E., Feinstein, H., Ouman, C.: Role-based access control model. IEEEComputer 29, 2(February). WIELEMAKER, J. SWI-Prolog (1996), http://www.swi.psy.uva.nl/projects/SWI-Prolog/
20. Wang, H., Cao, J., Zhang, Y., Varadharajan, V.: Achieving Secure and Flexible M-Services Through Tickets. In: Benatallah, B., Maamar, Z. (eds.) IEEE Transactions Special issue on M-Services (2003); IEEE Transactions on Systems, Man, and Cybernetics. Part A (IEEE 2003) 33(6), 697–708
21. Wang, H., Sun, L., Zhang, Y., Cao, J.: Authorization Algorithms for the Mobility of User-Role Relationship. In: Proceedings of the 28th Australasian Computer Science Conference (ACSC 2005), pp. 167–176. Australian Computer Society (2005)
22. Wang, H., Cao, J., Zhang, Y.: A flexible payment scheme and its permission-role assignment. In: Proceedings of the 26th Australasian Computer Science Conference (ACSC 2003), Adelaide, Australia, vol. 25(1), pp. 189–198 (2003)

23. Wang, H., Cao, J., Zhang, Y.: Formal authorization approaches for permission-role assignment using relational algebra operations. In: Proceedings of the 14th Australasian Database Conference (ADC 2003), Adelaide, Australia, February 2-7, 2003, vol. 25(1), pp. 125–134 (2003)
24. Wang, H., Cao, J., Zhang, Y.: Formal Authorization Allocation Approaches for Role-Based Access Control Based on Relational Algebra Operations. In: 3nd International Conference on Web Information Systems Engineering (WISE 2002), Singapore, pp. 301–312 (2002)
25. Wang, H., Cao, J., Zhang, Y.: A Consumer Anonymity Scalable Payment Scheme with Role Based Access Control. In: Proceedings of the 2nd International Conference on Web Information Systems Engineering (WISE 2001), Kyoto, Japan, pp. 73–72 (2001)
26. Yao, W., Moody, K., Bacon, J.: A model of OASIS role-based access control and its support for active security. In: Proceedings of ACM Symposium on Access Control Models and Technologies (SACMAT), Chantilly, VA, pp. 171–181 (2001)
27. Zhang, L., Ahn, G., Chu, B.: A Rule-based framework for role-based delegation. In: Proceedings of ACM Symposium on Access Control Models and Technologies (SACMAT 2001), Chantilly, VA, May 3-4, 2001, pp. 153–162 (2001)
28. Zhang, L., Ahn, G., Chu, B.: A role-based delegation framework for healthcare information systems. In: Proceedings of ACM Symposium on Access Control Models and Technologies (SACMAT 2002), June 3-4, 2002, pp. 125–134 (2002)

# Relevance Feedback Learning for Web Image Retrieval Using Soft Support Vector Machine*

Yifei Zhang, Daling Wang, and Ge Yu

School of Information Science and Engineering
Northeastern University, Shenyang 110004, P.R. China
{zhangyifei,dlwang,yuge}@mail.neu.edu.cn

**Abstract.** Eliminating semantic gaps is important for image retrieving and annotating in content based image retrieval (CBIR), especially under web context. In this paper, a relevance feedback learning approach is proposed for web image retrieval, by using soft support vector machine (Soft-SVM). An active learning process is introduced to Soft-SVM based on a novel sampling rule. The algorithm extends the conventional SVM by using a loose factor to make the decision plane partial to the uncertain data and reduce the learning risk. To minimize the overall cost, a new feedback model and an acceleration scheme are applied to the learning system for reducing the cost of data collection and improving the classifier accuracy. The algorithm can improve the performance of image retrieving effectively.

**Keywords:** Image retrieval, Relevance feedback learning, Soft support vector machine, Sampling rule.

## 1 Introduction

With the development of digit imaging and multimedia technology, image has become a major source of information in the Internet, which has attracted substantial research interest in providing efficient image indexing and search tools. Many existing image research engines are most based on text query and retrieve images by matching keywords from image filenames, ALT-tags and context in web pages, such as Google, Lycos, AltaVista and Yahoo [1, 2]. Although these engines can obtain some valid recalls, retrieval results are always short of precision because text information is usually rarely reliable or even irrelevant to the image content.

To integrate image content into image retrieval, Content based image retrieval (CBIR) has been widely explored to search visually similar images in the last decade [3]. In these systems, rather than describing images using text, the low-level visual features (color, texture, shape, etc.) are extracted to represent the images, such as QBIC [4], Photobook [5], Pictoseek [6], etc. However, the low level features may not accurately express the high level semantic concepts of images. To narrow down the

---

* This work is partially supported by the National Basic Research Program of China (2006CB303103) and the National Natural Science Foundation of China (60573090).

Y. Ishikawa et al. (Eds.): APWeb 2008 Workshops, LNCS 4977, pp. 201–209, 2008.

semantic gap, the relevance feedback is introduced into CBIR [7]. In addition, the relevance feedback may properly give expression to users' subjectivity and effectively clue on retrieval concepts.

In a relevance feedback driven CBIR system, after a user submits a query, similar retrieved images will be returned to users for labeling the relevance to the query concept. Then the system refines the results by learning the feedback from users. The relevance feedback procedure is repeated again and again until the targets are found. In many of the current relevance feedback learning systems, the users are always required to offer own relevance judgments on the top images returned by the system. Then the classifier is trained by the labeled images to classify images that are not labeled through matching the query concept. It belongs to passive learning. However, the top returned images may generally be the most similar to the object concept, but not the most informative ones. In the worst case, all the top images labeled by the user may be positive and thus the standard classification techniques can not be applied due to the lack of negative examples. Unlike the standard classification problems where the labels of data are given in advance, in relevance feedback learning the image retrieval system need to actively select the images to label. Thus it is natural to introduce active learning scheme into image retrieval. In this scheme the system return the images with the most information to solicit user for feedback, and display the most similar images at the last time. Data with the most information always are most uncertain to their classification. Labeling these data benefit constructing of the classifier by the greatest degree and improve feedback efficiency.

Being an existing active learning algorithm, Support Vector Machine (SVM) active learning has attracted the most research interest [8]. In the SVM classifier, the closer to the SVM boundary an image is, the less reliable its classification is. SVM active learning selects those unlabeled images closest to the boundary to ask users to label so as to achieve maximal refinement on the hyperplane between the two classes. Actually the convergent speed usually is slow using this strategy, and sometimes the effect is not so mach as using the most similar strategy. Otherwise, the major disadvantage of SVM active learning is that the estimated boundary may not be accurate enough by maximizing the margin space through iterative heuristic optimization.

Benefit from recent progresses on Support Vector Machine and semi-supervised learning [9, 10], in this paper we propose a novel active learning algorithm for image retrieval, called Active Soft SVM Learning (ASSVM). Unlike traditional SVM methods which aim to find a separating hyper-plane with the maximal margin, the separating hyperplane of Soft SVM is defined partial to uncertain points to decrease learning risk. Specifically, we introduce a new sampling rule into the relevance feedback learning. The new sampling rule aims to accelerate the classifier convergence by selecting the most informative data points which are presented to the user for labeling. Since the task of active learning is to minimize an overall cost, which depends both on the classifier accuracy and the cost of data collection, in the proposed framework web users does not need to label all of the returned images, but to select the most similar image to the retrieval concept. By contrast computing to the selected image, the system automatically gives feedback to active learning. Thus users' additional burden due to the interaction is highly alleviated.

The rest of the paper is organized as follows. In section 2, we provide a brief description of the related work. Our proposed Active Soft SVM Learning algorithm is introduced in section 3. Section 4 shows the application of ASSVM in relevance feedback image retrieval. Finally, we give some conclusions in Section 5.

## 2 Related Work

In this section, we give a brief description of the related work for our proposed algorithm, which includes the active learning problem and the basic framework of SVM relevance feedback.

### 2.1 The Active Learning Problem

The active learning problem is that the system as the learner returns the most informative data to the use as the teacher and the user make judgment for relevance feedback. The description of the active learning problem is the following. Given a set of points $X = \{x_1, x_2, ..., x_n\}$ in the vector space $R^d$, a subset of $X$, $Y = \{y_1, y_2, ..., y_m\}$, is found which contains the most informative points. If the user labels the points $y_i (i = 1, 2, ...m)$ as training data, the classifier is to be improved by learning these training data.

### 2.2 The SVM Relevance Feedback

The basic idea of SVM is to map data into a high dimensional space and find a separating hyperplane with the maximal margin, which separates the positive from the negative samples [11, 12]. Given training vectors $x_i \in R^n, i = 1, 2, ..., l$ in two classes, and a vector of label $y_i \in [-1, 1]$, SVM is equivalent to solve a quadratic optimization problem:

$$\min_{w, b, \xi, \rho} \frac{1}{2} w^T w + \frac{1}{l} \sum_i \zeta_i - v\rho \qquad (1)$$

$$\text{subject to} \quad y_i (w^T \phi(x_i) + b) \geq \rho - \zeta_i,$$
$$\zeta_i \geq 0, \rho \geq 0, i = 1, 2, ...l$$

where training data are mapped to a higher dimensional space by the function $\phi$, and $v(0 \leq v \leq 1)$ is a penalty parameter on the training error, and the positive error variable $\zeta_i$ is introduced for counting the error in nonseparable cases [13]. Thus it can be seen that the decision function $f(x) = \text{sgn}(w^T \phi(x) + b)$ is fairly impacted by the treated training samples.

As the SVM is used for relevance feedback learning of image retrieval system, its approach is generally started with a set of original returned images by text retrieval or low-level features retrieval. The user labels the returned images according their relevance information, where the positive samples marked as $I^+$ are relevant and the negative samples marked as $I^-$ are irrelevant. As the feedback from users, these labeled images are used as the train dataset inputted into the SVM to refine the classifier. Then the obtained classifier is applied to generate new retrieved results to users for the next feedback. The relevance feedback procedure is repeated again and again until the cease criteria is satisfied. Finally the system returns the top k images which are in the relevant area and furthest to the separating hyper-plane.

### 2.3  Version Space

In the machine learning problem, effectively shrinking the version space can accelerate convergence of the learner to the target concept. Since a small version space will guarantee that the predicted hyperplane is close to the optimal one, the goal for designing an active learning is to maximally shrink the version space after each selected sample is labeled by users. The version space of SVM is

$$V = \{ w \in W \mid \|w\| = 1, y_k (w^T \cdot \phi(x_x) + b) > 0, k = 1,2,..., k \} \quad (2)$$

where the parameter space $W$ is equal to the feature space $F$.

In literature [14], a criterion for selecting samples in active SVM (SVM$_{active}$) is given, i.e.

$$C_{SVM}(x_k) = \left| Area(V^+(x_k)) - Area(V^-(x_k)) \right| \quad (3)$$

Here, $Area(V^+(x_k))$ and $Area(V^-(x_k))$ denote the version space after the $i$th query $x_k$ is labeled as positive sample and negative sample respectively. SVM$_{active}$ selects the sample with the smallest $C_{SVM}$ in each round of relevance feedback to guarantee that SVM$_{active}$ maximally shrink the version space in Equation 2. Since it is practically difficult to compute the size of the version space, the method finally takes on the closest-to-boundary criterion to select examples [15, 16]. However, when the set of all conditional distributions of the samples is fixed, the SVM often has slow convergence to the target concept. J He et al. replaces this criterion by mean version space [17]:

$$C_{SVM}(x_k) = m^+(x_k)P(y_x = 1 \mid x_k) + m^-(x_k)P(y_k = -1 \mid x_k) \quad (4)$$

Here, $P(y_x = 1 \mid x_k)$ and $P(y_k = -1 \mid x_k)$ denote the posterior probabilities after an unlabeled example $x_k$ has been label, and $m^+(x_k)$ and $m^-(x_k)$ indicate the margins obtained after SVM is retrained when $x_k$ is added to positive example set and negative example set respectively. According this criterion, we should select samples which will maximally reduce the margin of separating hyperplane in each round of relevance feedback. In our method, Mean Version space is used to accelerate the convergence of Soft SVM.

## 3  ASSVM Algorithm

Regular SVM technology tries to refine the classifier by assuming that all of labels for training samples are trusted. However, if some labels of data are uncertain or fuzzy, these training data will have few effect on the predicted decision. Usually we give the label confidence degree of data according to relevance with the object concept instead of certain labels. Referring to the SLSVM algorithm in the literature [18], we propose ASSVM algorithm to take on classification by training the unconfident data. It is described in the subsequent paragraph.

In order to reduce the learning risk from fuzzy data, we introduce the loose factor $s_i$ to margin error $\zeta_i$ in equation 1. Let $s_i$ describe the relevance degree to the object concept, namely $s_i \in [0,1]$. Given the training data set $\{(x_1, s_1), (x_2, s_2), ..., (x_l, s_l)\} \subseteq X \times S, S = \{s_i \mid s_i \in [0,1], i = 1,2,..., l\}$, equation 1 is modified as:

$$\min_{w,b,\xi,\rho} \frac{1}{2} w^T w + \frac{1}{l} \sum_i s_i \zeta_i - v\rho \tag{5}$$

subject to   $y_i(w^T \phi(x_i) + b) \geq s_i \rho - \zeta_i,$
$$\zeta_i \geq 0, \rho \geq 0, 0 \leq v \leq 1, i = 1,2,...l$$

From the function, it can be seen that the smaller $s_i$ is, the larger the margin error $\zeta_i$ is, whereas the margin error $\zeta_i$ will become larger if $s_i$ is larger. In the other words, when the label confidence degree of data is small, the allowed margin error of separating hyperplane is large facing a non-separable example.

We transform the function to Lagrange equation for quadratic program:

$$L(w, \zeta, b, \rho, \alpha, \beta, \delta) = \frac{1}{2} w^T w + \frac{1}{l} \sum_i s_i \zeta_i - v\rho$$
$$- \sum_i (\alpha_i (y_i (w^T \phi(x_i) + b) - s_i \rho + \zeta_i) - \beta_i \zeta_i) - \delta\rho \tag{6}$$

We obtain the primal optimization result for Lagrange dual problem by taking the partial derivative of L with respect to $w, \zeta_i, b, \rho$ respectively, i.e.

$$\min_\alpha \frac{1}{2} \sum_{i,j} \alpha_i \alpha_j y_i y_j k(x_i, x_j) \tag{7}$$

subject to    $\sum_i \alpha_i y_i = 0,$

$$0 \leq \alpha_i \leq s_i \frac{1}{l}, \ i = 1,2,..., l,$$

$$\sum_i \alpha_i s_i \geq v.$$

The decision function is gained as the form below:

$$f(x) = \text{sgn}(\sum_i \alpha_i y_i k(x_i, x_j) + b) \tag{8}$$

Here, since the constraint of the dual variable $\alpha_i$ is changed from $0 \leq \alpha_i \leq \frac{1}{l}$ in regular SVM to $0 \leq \alpha_i \leq s_i \frac{1}{l}$ in ASSVM, the predicted result will be impacted by the label confidence degree of support vectors. The affect will be greater when the loose factor of the support vector is larger, and the smaller its loose factor is, the less the affect on the predicted result is.

## 4   Web Image Retrieval Using ASSVM

In this section, we describe how to apply ASSVM to CBIR. We firstly begin with a brief description of image representation using low-level visual features.

### 4.1  Visual Feature Representation

Image feature representation is a crucial part to CBIR. Three kinds of visual features, color, shape and texture, usually are used to describe images. Color and texture features are extensively used in CBIR due to their simplicity and effectivity. Shape features are usually extracted after images have been segmented into regions or objects. Since it is difficult to accurately segment images, shape features used for image retrieval has been limited to special applications where objects or regions are readily available. Generally edge features of the image are used in CBIR instead of shape features.

We choose three feature types as image low-level representation, color histogram, edge direction histogram and texture histogram. The color histogram is calculated using 4×4×4 in HSV space and form 64-dimension color feature vector. Edge direction feature we used is derived from the corresponding gray image obtained by transforming the color image. Then we transform the gray image to the edge image by morphological gradient algorithm. We describe the edge shape in each 8*8 subregion by 6 kinds of direction descriptor including empty, 0°, 45°, 90°, 135°, and confusion. Last we obtain a 6-dimension edge feature from the edge direction histogram to describe the edge information of the image. The image's gray co-occurrence matrix and the image's Gabor wavelet features are integrated for the image's texture expression. The gray co-occurrence matrix of image pixels is constructed from the gray image corresponding to the color image and the 16-dimension statistical feature is derived. At the same time, we perform the Discrete Wavelet Transformation (GWT) on the gray images employing a Gabor multi-resolution wavelet filter [19]. We construct 8 wavelet filters from 2 scales in 4 orientations. Then mean and variance of every subband after filtering are computed as wavelet features of the image and a 16-dimension wavelet-based texture feature is obtained. Combining the gray texture features and the wavelet texture features, we use a 32-dimension texture vector to describe the texture information for each image.

## 4.2   The Training Dataset

The training dataset for ASSVM learning is derived from the use's feedback data. It is well known that good training dataset can accelerate convergence of the classifier. Key of the classifier construction is what images will be selected to request the user for labeling.

SVM active learning usually selects those unlabeled images closest to the boundary to ask users to label as Fig.1. The images are considered with the most classification information compared with the most relevant images. Considering the acceleration scheme proposed in the section 2.3, we need to build a good version space using equation 4 through one specified relevant example. Since the most relevant images always lie in the neighborhood of the positive examples, we apply a novel sample selection rule for ASSVM as following: the images lying in the margin area and closest to the relevant boundary are returned to the user for labeling. The rule is specified in Fig.2.

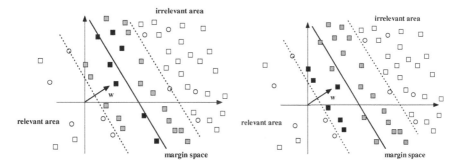

**Fig. 1.** Sampling closest to the boundary    **Fig. 2.** Sampling closest to the relevance area

As the loose factor expressing the label confidence degree is introduced in ASSVM, the user does not need to label all of the returned images from the previous round, but only to select the most relevant image for feedback. Then the system computes the similarity of every returned unlabeled image with the feedback image as loose factor for ASSVM. The training dataset is obtained by combining the feature vector of every image and its loose factor.

## 4.3   Relevance Feedback Image Retrieval

Relevance feedback is one of the most important techniques to narrow down the semantic gap between high-level semantic concepts and low-level visual features [7]. The user's relevance feedbacks are generally used to improve the sample dataset or adjust the vector weighting. It is an on-line learning method in which the image retrieval system acts as a learner and the user acts as a teacher. ASSVM Web image retrieval approach is taken on as follows:

(1) The system ranks the returned images from the existing image retrieval engine according to pre-defined distance metric and presents to the user the top ranked images.

(2) The user selects one image most relevant to the query concept for feedback. The system constructs the original classifier using the training dataset through combining the image low-level features and the label confidence degree gained by similarity computing.

(3) The system returns to the user the top images according to the new sampling rule described in the previous section. The user selected the most relevant image from returned images and the training dataset is created according to the user's feedback.

(4) The system refines the classifier by learning the training data in the version space which is constructed by equation 4 when $x_k$ is specified as the feedback image.

Go to step 3 until the stop criterion is satisfied.

(5) The system returns the most relevant images generated by the classifier.

The approach may be used in multi-user pattern too. If some users carry out feedback simultaneously, different images would be like to choose as most relevant images. We compute the label confidence degree in respect to every most relevant and select the maximum confidence degree of each image to be its loose factor. The version space is confirmed by choosing the max of $C_{SVM}$ (in equation 4) derived from every selected image. Thus we can considerate the subjective comprehension of every user to the query concept.

## 5  Conclusion

We proposed a novel method of active learning method for Web image retrieval called AASVM. Our algorithm introduced a loose factor expressing label confidence degree in SVM for image relevant feedback and proposed a novel sampling rule considering the trade-off between the most informative feedback and the acceleration scheme for convergence of the learner. The use of the loose factor enhances the impact of the uncertain data to the predicted results and reduces the learning risk. In addition, only one image feedback greatly decreases the user's burden due to the interaction.

## References

1. Kherfi, M.L., Ziou, D., Bernardi, A.: Image Retrieval From the World Wide Web: Issues, Techniques, and Systems. ACM Computing Surveys 36(1), 35–67 (2004)
2. Shen, H.T., Ooi, B.C., Tan, K.-L.: Giving Meaning to WWW Images. In: ACM Multimedia, LA, USA, pp. 39–47 (2000)
3. Smeulders, A.W., Worring, M., Santini, S., Gupta, A., Jain, R.: Content-based image retrieval at the end of the early years. IEEE Transactions on Pattern Analysis and Machine Intelligence 22(12), 1349–1380 (2000)
4. Niblack, W., Barber, R., et al.: The QBIC project: Querying images by content using color, texture and shape. In: Proc. SPIE Storage and Retrieval for Image and Video Databases (February 1994)
5. Pentland, A., Picard, R.W., Sclaroff, S.: Photobook: Content-Based Manipulation of Image Databases. Intl. J. Computer Vision 18(3), 233–254 (1996)

6. Gevers, T., Smeulders, A.W.M.: Pictoseek: Combining Color and Shape Invariant Features for Image Retrieval. IEEE Trans. Image Processing 9(1), 102–119 (2000)
7. Rui, Y., Huang, T.S., Ortega, M., Mehrotra, S.: Relevance feedback: A power tool for interative content-based image retrieval. IEEE Transactions on Circuits and Systems for Video Technology 8(5) (1998)
8. Tong, S., Chang, E.: Support vector machine active learning for image retrieval. In: Proceedings of the ninth ACM international conference on Multimedia, pp: 107–118 (2001)
9. Goh, K.-S., Chang, E.Y., Lai, W.-C.: Multimodal concept-dependent active learning for image retrieval. In: Proceedings of the ACM Conference on Multimedia, New York, USA (2004)
10. He, X.: Incremental semi-supervised subspace learning for image retrieval. In: Proceedings of the ACM Conference on Multimedia, New York, USA (2004)
11. Vapnik, V.N.: Statistical Learning Theory. Wiley, New York (1998)
12. Scholkopf, B., Platt, J.C., Shawe-Taylor, J., Smola, A.J., Williamson, R.C.: Estimating the support of a high-dimensional distribution. Neural Computation 13(7), 1443–1471 (2001)
13. Atkinson, A.C., Donev, A.N.: Optimum Experimental Designs. Oxford University Press, Oxford (2002)
14. Scholkof, B., Smola, A.J., Williamson, R., Bartlett, P.: New support vector algorithms. Neural Computation 12, 1083–1121 (2000)
15. Tong, S., Koller, D.: Support vector machine active learning with applications to text classification. Journal of Machine Learning Research 2, 45–66 (2001)
16. Tong, S., Chang, E.: Support vector machine active learning for image retrieval. In: Proc. 9th ACM Int. Conf. on Multimedia, Ottawa, Canada (2001)
17. He, J., Li, M., Zhang, H.J., et al.: Mean Version Space: a New Active Learning Method for Content-Based Image Retrieval. In: Proceedings of the 6th ACM SIGMM International Workshop on Multimedia Information Retrieval, New York, USA, pp. 15–22 (2004)
18. Hoi, C.-H., Lyu, M.R.: A Novel Log-based Relevance Feedback Technique in Content-based Image Retrieval. In: Proceedings of the 12th ACM International Conference on Multimedia, New York, USA, pp. 24–31 (2004)
19. Manjunath, B.S., Maw, Y.: Texture features for browsing and retrieval of image Data. IEEE Transaction on Pattern Analysis and Machine Intelligence 8(18), 837–842 (1996)

# Feature Matrix Extraction and Classification of XML Pages[*]

Hongcan Yan[1,2], Dianchuan Jin[2], Lihong Li[2], Baoxiang Liu[2], and Yanan Hao[3]

[1] School of Management, TianJin University, Tianjin 300072, China
[2] College of Sciences, HeBei Polytechnic University, Hebei Tangshan 063009, China
[3] School of Computer Science and Mathematics, Victoria University, Australia

**Abstract.** With the increasing data on the Web, the disadvantage of HTML is more and more evident. There must be a method which can separate data from display, and then XML (eXtensible Markup Language) arises. XML can be the main form of expressing and exchanging data. How to store, manage and use the data effectively have been problems needing to be solved in the field of Internet, in which the automatic text classification is an important one. In this article, we propose a data model to analyze documents using the hierarchical structure and keywords information. Experiments show the model has not only high accuracy, but also less time cost.

**Keywords:** XML Page Classification, Frequency Structure Hierarchy Vector Space Model, Combined Features Extracting, Rough set, Page Feature Matrix.

## 1 Introduction

With the increasing data on the Web, the disadvantage of HTML is more and more evident. There must be a method which can separate data from display, and then XML (eXtensible Markup Language) arises. It is easy to operate and convenient to fulfill in the circumstance of WWW, for it retains many advantages of SGML. Especially, its self-descriptive data structure can embody the relationship among data, and the data can be operated by applications more conveniently. Thus XML can be the main form of expressing and exchanging data. How to store, manage and use the data effectively have been problems needing to be solved in the field of Internet, in which the automatic text classification is an important one.

An XML document is a synthesis of textual contents and structural information. Its analysis is different from traditional methods, because of the way of gaining and using structure information. Traditional document classification methods do not work well when applied to XML documents classification. So it is necessary to study a new specialized method about XML text classification.

In recent years, although research have been paying more and more attention to the analysis and process of XML documents and semi-structured data, the achievements

---

[*] This work was supported by Natural Scientific Fund Project of Hebei Province (F2006000377), the Ph.D. Programs Foundation of Ministry of Education of China (No. 20020056047).

Y. Ishikawa et al. (Eds.): APWeb 2008 Workshops, LNCS 4977, pp. 210–219, 2008.

in text minding are relatively little. However, many people have pushed the research on semi-structured data classification forward. For instance, Yi proposed an expanded vector model for semi-structured documents classification. This model describes document elements using nested vectors, then classify documents by probability statistics[1]; Denoyer applied Beyesian network model to semi-structured document classification[2]; Zhang suggested calculating similarity of XML documents with edit distance[3]; Regarding the structural information as time series, Flesca computed structural similarities between XML documents by time series analysis[4]; Zaki and others put forward cost sensitive structured rules for mining frequent sub-trees. It has good effect on semi-structured data classification [5]. Although these ideas are new and original, they can not be applied into general cases.

In this article, we propose a data model to analyze documents using the hierarchical structure and keywords [9-11] information. The model combines the ideas of XRules (Structural rule-based classification) [5], mining algorithm for unordered frequent sub-trees [6] and N level vector space model [7-8]. Through constructing a decision table and using the superior reduction of rough sets, this model achieves the aim of reducing the count of dimensions of feature values and implements the classification based on rules. Experiments show the model has not only high accuracy, but also less time cost.

## 2  XML Document Model and Feature Analysis

XML documents are text files using Unicode format. Their basic contents include XML declaration, comment, tag, element, attribute and text. The front two parts mainly describe the edition information and uses of documents. The other four parts mainly reflect data which is the main content of our study. A valid XML document can be regarded as a node tree with label, which is called a document model. Figure 1 shows the XML document structure model of a journal and a conference, respectively.

**Fig. 1.** XML document structure model of a Journal and a Conference paper

A XML document is corresponding to a directed tree T=<V, E>, in which V is composed of all elements or attributes. And for the set of E, it is defined as: if element $a \in V$, and its sub-element or attribute $b \in V$, then $(a, b) \in E$. From figure1, we can see that the elements and sub-elements have different levels in a XML document and the node tags on different levels have different importance, so the classification should have different weight to assign. Actually the data features in documents usually appear in the nodes of document tree. We can only consider the leaf nodes.

# 3  Feature Matrix Combining Structure and Keywords in XML Documents

XRules classification only considers the structure information of Xml documents. The classification is implemented through finding the structural rules which can satisfy the certain support. Obviously, besides the structure influencing XML classification, the key words of the document is quite important. As an illustration, when two documentary materials are classified according to different fields of academic research, what we should do are both considering the structure of the document and paying more attention to dealing with the content it describes.

## 3.1  Frequent Structural Level Vector Space Model

In "Similarity Measures for XML documents Based on Frequency structural Vector Model", the author described the model in details, and developed it further, then got the vector space model of frequent structural levels. With this way, the model divides a document into N levels on structure and constructs the fitting vector about the text feature and the weight of its level. Actually, regarding each frequent structure as a vector is similar to N level text paragraph in VSM (Vector Space Model). So the whole XML document can be expressed as a group of vectors, shown by a matrix in table 1. Thus we can analyze both the structure and text content information of semi-structural text.

**Table 1.** The frequency structure hierarchy space feature value matrix

| Frequent structure feature \ Text level feature | $Key_1$ | $Key_2$ | ... | $key_m$ |
|---|---|---|---|---|
| $S_1$ | $d_1(1,1)$ | $D_1(1,2)$ | | $d_1(1,m)$ |
| $S_2$ | $d_2(2,1)$ | $d_2(2,2)$ | | $d_2(2,m)$ |
| ....... | | | | |
| $S_k$ | $d_k(k,1)$ | $d_k(k,2)$ | | $d_k(k,m)$ |

## 3.2  Calculating the Association Degree about Combination of the Structure and Content

We use TreeMiner+ algorithm [17] to mine the frequent structure in XML documents. The node structure unit can record the level attribute from the source of document. This relates to the weight of text. In the frequent structure of Figure 1(supposing Minsup=50%), the set of node {Journal, Name, Vol, Articles, Title, Author, Abstract, KeyWord, FullText, Conference, Resume} are all frequent nodes. In other words, they are 1-subtrees. There are eighteen 2-subtrees in total and what figure 2 shows is partial subtrees. The weight function [17] of each frequent structure is:

$$d_{si} = \delta_i(s_i) * B(s_i) * \log(|D| / |DF(s_i) + 0.5|) \qquad (1)$$

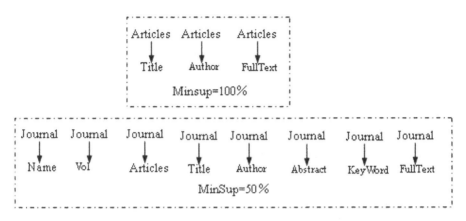

**Fig. 2.** The part of frequent 2-subtree

We number the leaf-node in each document model tree by the order of root-first, as shown figure 3. Then we calculate the frequency of keywords in leaf nodes respectively in the statistical way $TF_{k,d}^{j} = tf_{k}^{j}(d)$ , where $j$ is the number of its order. The weight value of the position of the keywords in the leaf node text from XML document can be calculated as $w_{k,d}^{j}$ = the level weight value of its parent node / the count of its sibling nodes (the level weight value of root is 1, and $\sum w_{k,d}^{j} = 1$ ).

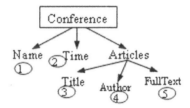

**Fig. 3.** Leaf node code of XML documents

For example, the position weight value of "Name" is 1/3, the position weight value of "Author" is 1/9. So in XML document d, the $K^{th}$ keyword's frequency can be shown as:

$$TF_{k,d}^{i} = \sum (w_{k,d}^{j} * TF_{k,d}^{j}) \qquad (2)$$

Combining the common text and HTML weight calculating method provided by Salton[18], we correct the formula of the association degree of each keyword as:

$$w_{i}(d) = \frac{TF_{k,d}^{[i]} \log(N / n_i)}{\sqrt{\sum_{j} (TF_{k,d}^{[i]} \log(N / n_i))^2}} \qquad (3)$$

In formula 3, N is the number of all documents and $n_i$ is the number of the documents having the keyword.

An XML document can be considered as a combination of structure and several keywords of the document, as shown by a matrix of k*m in table 1, where k is the number of different frequent structures (sorted on MinSup by descending), m is the number of different keywords, the elements of matrix $d_x(i,j)$ come from the frequency of keyword $w_j$ occurring in frequent structure $s_i$ of document $doc_x$.

## 4    Classification Method Based on Rough Set

After feature weight calculating, character selecting, and detecting the structure which is included totally appearing in frequent structure (shown in figure 4), we can ensure that there is no the repeated structure in frequent structure set. Then we have the matrix composed of feature value of structure and content in training XML documents, as shown in table 2, in which $S=\{S_1, S_2, \ldots \ldots S_k\}$ is a frequent structure which is totally different from others. Ds, weight of every frequent structure, is gotten from formula 1; $\{KW1, KW2, \ldots \ldots KWm\}$ are keywords about content characteristics. TFi(Sj) is the frequency of keyword i appearing in structure Sj, gotten from formula 2.

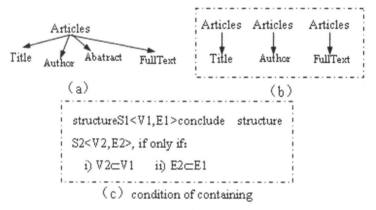

**Fig. 4.** Containment in frequency structure (a) contains (b)

**Table 2.** Frequency structure hierarchy vector space feature value

| S | Ds | KW1 | KW2 | ....... | KWm |
|---|----|-----|-----|---------|-----|
| S1 | ds1 | TF1(s1) | TF2(s1) | ...... | TFm(s1) |
| S2 | ds2 | TF1(s2) | TF2(s2) | ...... | TFm(s2) |
| ....... | ....... | ...... | ...... | ...... | ...... |
| Sk | dsk | TF1(sk) | TF2(sk) | ...... | TFm(sk) |

Developed by Department of Computer and Information Science, Norwegian University of Science and Technology, ROSETTA[19] is a comprehensive software system for conducting data analyses within the framework of rough set theory. The ROSETTA system boasts a set of flexible and powerful algorithms, and sets these in a user-friendly environment designed to support all phases of the discernibility-based modeling methodology. It can do well in attribute reduction, rule extracting, classification and comments. We adopt ROSETTA system to classify tested data set. Figure 5 is the detailed circuit of our processing.

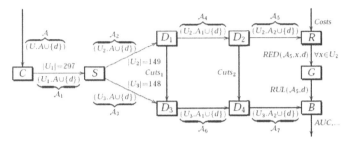

**Fig. 5.** ROSSETA system machine learn structure

The content of the input decision system $A$ is the feature value of tested document set ( the data in table 2). $A$ is first cleansed of missing values to produce $A_1$, and then split into two disjoint subsystems $A_2$ and $A_3$. $A_2$ is discretized in a two-stage procedure, and $A_3$ is discretized using the cuts computed from $A_2$. Reduction and rules are then computed from the processed version of $A_2$, and the rules are used to classify the objects in the processed version of $A_3$. From this classification, performance estimates are estimated.

### 4.1 Classification Based on Structure

Without considering about the content of the document and only remain feature value Ds column of frequent structure in table 2, we obtain the feature space of document which is composed of all frequent structures, and thus acts as the content of matrix $A$.

### 4.2 Classification Based on Content

Without considering about the structure of documents and only remaining the weight of keywords in table 2, we obtain the feature space of document which is composed of all keywords, and makes the content of matrix $A$ from formula 3.

### 4.3 Classification Based on the Combination of Structures and Contents

Sometimes classification of XML documents is influenced more by structure. For example, from the point of publication in a journal or a conference, document 2 from figure 1 and the document from figure 3 can be classified with the same way. They belong to the same class. But sometimes, considering about the content of articles, the feature keywords can play prominent roles, so an article maybe belong to multiple classes. In order to get different classification, we can give different weights to structures and contents and extract feature value of the combination of structures and contents.

Every case about the feature of keywords in frequent structures is an independent event, and the case between every frequent structure and the keywords in it is an apriori event. In table 2, every keyword KWi's weight TFi to the whole document not only related to its level, but also with its weight in the frequent structure. So the formula will be like this:

$$W_i(d) = \sum TF_k(S_k) * d_{sk}(k = 1, 2, ..., n)$$ (4)

In document d, the weight of the i$^{th}$ keyword is the accumulation of the frequency of every frequent structure multiplied by its weight. So the frequent structure level vector space is organized as a vector from a matrix by this way, and the combined feature of structures and contents in whole documents is set into a matrix composed of the weight of m keywords, which is the content of $\mathcal{A}$.

## 5   Results and Analysis of Experiments

All experiments were performed on a 2.10GHz AMD Athlon 4000+ PC with 1G memory running Windows 2000 Server. The algorithm is performed by JAVA programming with Java 2 Platform Standard Edition 5.0.

There are two kinds of experimental data. The first one is to use OrdinaryIssuePage and IndexTermsPageXML of ACMSIGMOD[20] to test the classification based on structures of documents. Table 3 shows the sub-sets of data. The first number of Total num. of documents is used to train frequent sub-tree mining, and the next one is used to test classification. There is some information about the classification in ACMSIG-MOD documents. Every file usually belongs to multi-classes. If a document feature satisfies the classification rules of many files, then we can conclude that it belongs to multiple classes. During the experiments we use classified information instead of artificial mark. From the formula 5, the precision of the experiment results can be 98.7%.

P = *the number of documents classified correctly / the number of all tested*    (5)
*documents*

**Table 3.** Sub-set of data in experiments

| Datasets | Sources | Total num. of documents |
|---|---|---|
| ACMSIGMOD-1 | OrdinaryIssuePage(1999) | 40+20 |
| ACMSIGMOD-2 | OrdinaryIssuePage(2002) | 20+10 |
| ACMSIGMOD-3 | IndexTermsPage(1999) | 40+20 |

The other data is SINA net news based on XML schema. These unsophisticated RSS news are provided by http://rss.sins.com.cn. Figure 6(a) shows the master document structure tree of every news page; figure 6(b) shows the structure of every news item. In order to illustrate how structures and contents influence classification, we modify somewhere on structures from XML net files, for example, adding or deleting some unimportant elements (iamge, link, copyright, guid and so on）.We mainly test combined structure and content feature extracted from news item. To simplify the experiment process, we finish it with artificial way to collect the content feature.

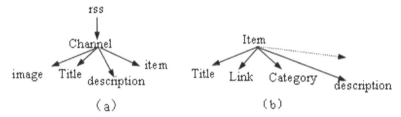

**Fig. 6.** Main structure elements of news net

We picked up 260 sheets of XML net pages. Among them 180 sheets are treated as training sets, divided into finance and economics, military, science and technology, automobile, physical culture. The comparison is done through two ways: (1) considering about frequency of keywords feature for each document; (2) considering about not only frequent structure feature, but also the position weight and frequency of keywords.

After obtaining feature value of matrix $A$, we train and test it with ROSSETA system. We show the result from in table 4. The precision and recall ration can be derived from formula 5 and formula 6.

$$R= （the\ number\ of\ documents\ classified\ correctly） /（the\ number\ of\ all\ the\quad (6)$$
$$original\ documents）$$

**Table 4.** Classification Performance Based on XML Structure and Keywords Frequency

| Dataset | | Economic News | Military News | Automobile News | Technical News | Sports News |
|---|---|---|---|---|---|---|
| TrainingSamples（1580） | | 30 | 30 | 30 | 30 | 30 |
| TestingSamples（80） | | 10 | 10 | 10 | 10 | 10 |
| Frequency structures（56） | | 13 | 11 | 9 | 12 | 11 |
| Keywords（232） | | 62 | 44 | 25 | 65 | 36 |
| Recall (%) | Only Keywords | 75.6 | 76.8 | 85.2 | 72.3 | 85 |
| | Considering Structure | 91.5 | 93.1 | 95.8 | 92.3 | 96.2 |
| Precision (%) | Only Keywords | 83.2 | 86.5 | 87.4 | 84.6 | 87.6 |
| | Considering Structure | 95.6 | 97.2 | 97.6 | 96.3 | 98.1 |

From the experiments result, we can see that in XML document classification, by considering about content and structure of document at the same time, then extracting combined structure and keywords feature value, we can have higher precision and recall ratio for classification.

## 6   Conclusions

In this article, we use the hierarchical structure and keywords information for XML documents classification. By constructing a decision table and using the superior reduction of rough sets, we can reduce the count of dimensions of feature values and implement the classification based on rules. Experiments show the model has high accuracy and less time cost.

## References

1. Yi, J., Sundarcsan, N.: A classifier for semi-structured documents. In: Ramakrishnan, R., Stolfo, S., Pregibon, D. (eds.) Proc. of the 6th ACM SIGKDD Int'l Conf. on Knowledge Discovery and Data Mining (KDD 2000), pp. 340–344. ACM Press, New York (2000)
2. Denoyer, L., Gallinari, P.: Bayesian network model for semi-structured document classification. Information Processing and Mangement 40(5), 807–827 (2004)
3. Zhang, Z.P., Li, R., Cao, S.L., Zhu, Y.: Similarity metric for XML documents. In: Ralph, B., Martin, S. (eds.) Proc. of the 2003 Workshop on Knowledge and Experience Management (FGWN 2003), Karlsruhe, pp. 255–261 (2003)
4. Flesca, S., Manco, G., Masciari, E., et al.: Detecting structural similarities between XML documents[A]. In: Proc. 5th Int. Workshop on the Web and Databases(WebDB 2002) [C], Madison,Wisconsin (2002)
5. Zaki, M.J., Aggarwal, C.C.: Xrules: An effective structural classification for XML data [C]. In: Int'l Conf. on Knowledge Discovery and Data Mining (SIGKDD 2003), Washington, DC (2003)
6. Ma, H.-B., Wang, L.: Efficiently Mining Unordered Frequent Trees. Mini-Macro Systems[J]
7. Ran, Z., Moydin, K.: Web Information Retrieval Based on XML and N -level VSM. Computer Technology and Development 16(5) (May 2006)
8. Niu, Q., Wang, Z.-x., Chen, D., Xia, S.-x.: Study on Chinese web page classification based on SVM. Computer Engineering and Design 28(8) (April 2007)
9. Yuan, J.-z., Xu, D., Bao, H.: An Efficient XML Documents Classification Method Based on Structure and Keywords Frequency. Journal of Computer Research and Development
10. Yang, Y.-C., Yang, B.-R., Zhang, K.-J.: Study of Rough Set Text Categorization Techniques Based on Combined Extracting Features. Application Research of Computer
11. Tang, K.: Method of classification based on content and hierarchical structure for XML file. Computer Engineering and Applications 43(3), 168–172 (2007)
12. Han, J.-T., Lu, Z.-J., Qin, Z.: The Research on Automatic ClassificatiOn Method of Complex Information System Based on XML. Systems engineering theory methodology applications 14(6) (December 2005)
13. Qu, B.-B., Lu, Y.-S.: Fast Attribute Reduction Algorithm Based on Rough Sets. Computer Engineering 33(11) (2007)

14. Wang, X., Bai, R.: Web Document Classification Method Based on Variable Precision Rough Set Model
15. Chen, W.-L., Chen, Z.-N., Bin, H.-Z.: Intelligent Search of the Process Web Pages Based on Rough Sets. Journal of wuhan university of technology 27(11) (November 2005)
16. Li, T., Wang, J.-P., Xu, Y.: A Rough Set Method for Web Classification. Mini –Micro System 24(3) (March 2003)
17. Yan, H.-C., Li, M.-Q.: Similarity Measures for XML documents Based on Frequency structure Vector Model
18. Salton, G.: Introduction to Modem Information Retrieval [M]. McGraw Hill Book Company, New York (1983)
19. Øhrn, A.: Discernibility and Rough Sets in Medicine: Tools and Applications. Department of Computer and Information Science Norwegian University of Science and Technology. N-7491 Trondheim, Norway
20. `http://www.acm.org/sigs/sigmod/record/`
    `XMLSigmodRecordMarch1999.zip`
    `http://www.acm.org/sigs/sigmod/record/`
    `XMLSigmodRecordNov2002.zip`

# Tuning the Cardinality of Skyline

Jianmei Huang, Dabin Ding, Guoren Wang, and Junchang Xin

Northeastern University, Shenyang 110004, China
{wanggr,xinjunchang}@mail.neu.edu.cn

**Abstract.** Skyline query has its own advantages and are useful in many multi-criteria decision support applications. But for the rigidity of skyline dominance relationship, the cardinality of skyline result cannot be controlled, either too big or too small to satisfy users' requirements. By relaxing the dominance relationship to k-skyline or general skyline, we propose a unified approach to find a given number of skyline. We call our output of skyline as $\delta$-skyline, in which $\delta$ indicates the number of skyline result. Without any user interference such as assigned weights or scoring functions, we are the first to propose a method to tune the cardinality of skyline operator in both directions, to either increase or decrease according to the requirement of user. To tune the cardinality of skyline, we adopt the concept of k-dominate and also we propose a new concept of general skyline. A point p is in general skyline if p is skyline in some subspace. General skyline have their meaning for they are the best at some aspects and are good alternatives to fullspace skyline. Finally, we present two algorithms to compute $\delta$-skyline. Extensive experiments are conducted to examine the effectiveness and efficiency of the proposed algorithms on both synthetic and real data sets.

## 1  Introduction

Skyline is a multicriteria decision support query and is quite important for several applications. Given a set of data points S, skyline query will return those points that cannot be dominated by any other data points. Here data point p dominate point q means that p is better than or equal to q on every dimension of data points and better than q on at least one dimension. A classic example of skyline query is to find hotels in Nassau, a seashore city in Bahamas, that are cheap and near to the beach. In Figure 1, a point in the figure represent a hotel, and has two attributes: price and distance to the beach. The skyline consists of points in the dashed line. We can infer that for different preference of different customers, the result that a customer want is in the skyline results.

Compared with Top-k query, skyline has some advantages: first it does not depend on a fixed scoring function which may not always be appropriate. Second, scales of different dimensions do not affect the skyline result. Although skyline is a powerful technique to retrieve interesting objects, it has its adherent problems. (1) The cardinality of skyline operator cannot be controlled. It is only determined by the intrinsic character of data and usually increases exponentially as the number of dimensions increase. (2) There are no ranks of the skyline results.

Y. Ishikawa et al. (Eds.): APWeb 2008 Workshops, LNCS 4977, pp. 220–231, 2008.

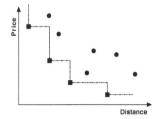

| Dimension | Skyline cardinality | Candidates |
|-----------|---------------------|------------|
| 17 | 1172 | 1346 |
| 10 | 428 | 1298 |
| 5 | 42 | 593 |
| 4 | 22 | 469 |
| 3 | 11 | 38 |
| 2 | 4 | 20 |

**Fig. 1.** An example of skyline          **Fig. 2.** Skyline of NBA players

It does not reflect users' preference at different dimensions and may lost its meaning when skyline size is big. The phenomena that the size of skyline increase rapidly as the number of dimensions increase is called curse of dimensionality. For example, Figure 2 shows the result of skyline on NBA data set, which includes statistics of 17300 player with 17 attributes. When the dimension is more than 10, the skyline cardinality is more 400, which may be overwhelming and lost its meaning. While several methods has been proposed to address this problem such as k-dominate in [7] and k representative skyline in [4], but no researchers consider a different scenario: suppose that a customer decide to buy a new mobile phone and launch a query in the database on the dimensions of memory size and screen size. Specifically, the customer values that the bigger the size of memory and screen are, the phone is the better. Assume that skyline operator retrieves only two best phones, but the customer may need more candidates(let us say the number is k) to compare with and make his or her choice or to buy as a substitute when the best two kinds of phone are not available. In a word, the current skyline operator cannot meet the need of skyline query if user expects fixed number of result.

Motivated by these problems, in this paper we propose a unified approach overcoming the drawbacks of the conventional skylines and tuning the cardinality of skyline operator in both directions. In particular, we define the concept of general skyline, which include all the subspace skylines. We show that general skyline is a meaningful extending of full space skyline when full space skyline is not sufficient. Also by adopting the concept of k-dominate to lower the cardinality of skyline, we propose two different algorithms to answer the skyline query at a given number of result. In summary,the contributions of this paper are the following:

- We define the concept of general skyline, which include all the subspace skylines. We show that general skyline is a meaningful extending of skyline operator when cardinality of skyline operator is not sufficient.
- To the best of our knowledge, without any user interference such as assigned weights or scoring functions, we are the first to propose a method to tune the cardinality of skyline operator in both directions, to either increase or decrease according to the requirement of user. So the advantage of skyline over top-k query can be maintained.

- We proposed two algorithms to achieve the goal of tuning the cardinality of skyline operator. Our experiments on both the synthetic and real NBA data verify the effectiveness of our algorithms.

The rest of this paper is organized as the following. Section 2 briefly reviews the related work. Section 3 gives some preliminaries and introduce the definition of general skyline. Section 4 proves some theorems and gives out our unified approach toward our aim. The extensive experiments results to show the effectiveness of the proposed algorithm are reported in Section 4. Finally we conclude this paper in Section 5.

## 2    Related Work

Ever since Borzsonyi et al. in [1] first introduced maximum vector problem into relational database context, defined it as skyline and proposed several main memory algorithms to compute skyline, different aspects and variants of skyline have been researched. Yuan et al. proposed the concept of *skycube* [9], which consists of all possible subspace skyline. In online query processing systems, Lin et al. [18] developed algorithms to compute skyline over data streams. The concept of the *skyline frequency* is proposed in [12], which measures the number of subspaces that a point is a skyline point. Lin et al. in [4] studied the problem of selecting k skyline points so that the number of points, which are dominated by at least one of these k skyline points, is maximized. The *thick skyline* in [17] can return points that are near to real skylines, so that it can increase the number of returned skylines. However, these variants have their shortcomings because they only focus on decreasing or increasing the number of skyline objects. Next, we will describe two works that are similar to or relate with our problem.

### 2.1    k-Dominate

[7] proposed a concept of *k-dominate* to relax the idea of dominance to k-dominance. A point p is said to k-dominate another point q if there are k ($\leq$ d) dimensions in which p is better than or equal to q and is better in at least one of these k dimensions. A point that is not k-dominated by any other points is in the k-dominant skyline. They studied some properties of k-dominant skyline and by running on some real data sets, they proved that k-dominant skyline could retrieve super stars in NBA data set or find top movies in movie rating data set. In this paper, we will further exploit some proprieties of k-dominate and integrate it into our unified approach to tuning the cardinality of skyline operator.

### 2.2    $\varepsilon$-Skyline

Quite recently, Xia et al. [2] focused on the same problem as our concerns and proposed the concept of $\varepsilon$-skyline to overcome three drawbacks of skyline as it mentioned: 1) size of skyline could not be controlled; 2) skyline objects do not have built-in ranks; 3) skyline does not reflect user preference at different dimensions. In the definition of $\varepsilon$-skyline, the dominance relation are changed into $t_1.i*w_i < t_2.i*w_i+\varepsilon$, in which $w_i$ (between $[0,1]$) is the weight on dimension i, and $\varepsilon$ is a constant between $[-1,1]$. The authors proved that the $\varepsilon$-skyline is monotone when $\varepsilon$ varies from -1 to 1. By assigning weights on dimensions, they said that the skyline result could be ranked. In fact, the problem we want to address are slightly different with the problems in [2]. We would like to keep the advantages of skyline and do not require user interference to control the size of skyline result. In addition, how to rank the skyline result can be done after the skyline query finishes. Meanwhile, we found two problems in the $\varepsilon$-skyline. First, each dimension of data point have to be normalized to $[0,1]$, otherwise w and $\varepsilon$ would have no effect in the inequation $t_1[i] \cdot w_i \leq t_2[i] \cdot w_i + \varepsilon$. Second, $\varepsilon$-skyline may abandon some interesting points as we can see in Figure 3, p2' can $\varepsilon$-dominate p3 and vice versa, then both p2 and p3 is discarded from the results. In general case, if two skyline points are near to each other, they may both be abandoned for no good reasons.

## 3  Preliminary and Definitions

Given a d-dimensional space D=$d_1,d_2,\ldots,d_d$ and a set of data points S, if p is a point in S, we use p.$d_i$ to denote the $i^{th}$ dimension value of point p. we us $\succ$ to denote the total order relationship(it could be $>, <$ or other relationship) on each dimension. Without loss of generality, we assume the relation is $>$ in the rest of this paper. Skyline query is defined as follows:

**Definition 1.** *(skyline) With points p,q in data set S, p is said to dominate q on space D, iff $\forall di \in D$, $p.d_i \geq q.d_i$ and $\exists p.d_t > q.d_t$. p dominates q is denoted as $p \succ q$ in this paper. Skyline is a set of points that are not dominated by any other points in data set S on space D.*

A subspace of D is a space that are composed by dimensions in D, specifically, any subspace $U \subseteq D$. It is easy to compute that the number of all subspace is $2^d$-1, and all the subspaces constitute a lattice as shown in figure 4. The skyline on D are usually called fullspace skyline, we can also compute skyline on any given subspace, which are called subspace skyline.

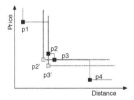

**Fig. 3.** An example of $\varepsilon$-skyline

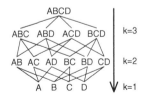

**Fig. 4.** An lattice example of skycube

### 3.1  δ-Skyline

The background of our problem is that a user first launches a skyline query. No matter how many data points are in the data set or the cardinality of dimension space is big or small, the user wants to set a limitation on the cardinality of the skyline results. The reason is quite very strong: first, if the skyline results are hundreds or thousands, then the results are overwhelming and the user can hardly get any interesting information from the results. So the user may have to adopt another technique to further reduce the size of results. Second, in the lower dimension space or if the distribution of the data points is correlated with each other, few predominating points may be retrieved. While in a sense that these points perfectly match the requirement of users, but for the rigidity of skyline by discarding all alternatives, they limits the choice of users. In real case, a top hotel may be full or disliked by users for some unmeasurable factors. Some other points that is the best in some aspects may be good candidates.

Although decreasing or increasing of the skyline cardinality can be done separately, but how to achieve it in a unified approach is a challenge, which is exactly what we are trying to solve.

**Definition 2.** *(δ-skyline) We define the output of our proposed algorithms as δ-skyline, in particular, δ is a number given by user.*

δ-skyline is different from top-k skyline for the following two reasons. First, there is no scoring function in the query, so there is no ranking of the query result. Second, δ-skyline may include some subspace skyline to increase the number of result. Next, we introduce the concept of k-dominate and for clarity, we define points that cannot be k-dominated are k-skyline points. The definition is the following.

**Definition 3.** *(k-skyline) A point p is said to k-dominate q $(d > k > 0)$, iff $\exists U \subseteq D$ ,$|U| = k, \forall ui \in U, p.ui \geq q.ui$ and $\exists ut \in U$, $p.ut > q.ut$. p k-dominate q is denoted as in this paper. k-skyline is a set of points that cannot be k-dominated any other points in data set S.*

### 3.2  General Skyline

If a point is the best on some aspects(dimensions), it is either a skyline or dominated by another skyline points. But this kind of points have meanings to users for their best quality on some aspects and could be good candidates if we are trying to find alternatives to skyline result.

**Definition 4.** *(general skyline)A point p is in general skyline iff there exists subspace U of fullspace D, p is a skyline on U.*

**Definition 5.** *(kg-gskyline) If a point p is a skyline on subspace U and $|U| = k$, then p is in k-general skyline. In short, We use kg-skyline to denote k-general skyline.*

## 4   Tuning Cardinality

In this section, we will first introduce and analyze some theorems that are the basis of our algorithms. Then, we propose two algorithms and explain in detail about how they can tune the cardinality of skyline operator in two directions.

**Theorem 1.** $|GeneralSkyline| \geq |skyline| \geq |k\text{-}skyline|$

Proof: (1) if p∈ General Skyline, then p∈skyline or p is dominated by some fullspace skyline. So $|GeneralSkyline| \geq |skyline|$. (2) if p,q∈ skyline, then $\exists k, k', p \underset{k}{\succ} q, q \underset{k'}{\succ} p$, so p∈ $k'$-skyline, q∈ $k$-skyline. so $|skyline| \geq |k - skyline|$.   □

Theorem 1 tells us that the cardinality of skyline is bigger than k-skyline and smaller than general skyline. So if the cardinality of skyline does not satisfy the user's need(not equal to $\delta$), we can move on to compute k-skyline to decrease the number of skyline or to compute general skyline to increase the number of skyline.

**Theorem 2.** $|(k+1)\text{-}skyline| \geq |k\text{-}skyline|$

This theorem was proved in the work of Chan [7], so we would not prove it. This theorem ensures that we can control the size of k-skyline by decreasing k to a proper quantity. But there is no such kind of property for general skyline. If $p \in$ k-gskyline, then it is possible that $p \notin (k+1)g$-skyline and $p \notin (k-1)g$-skyline. this can be proved by an enumerating an example. So the cardinality of general skyline is not monotone when k is decreasing compared with k-skyline. Although when k is decreasing, the kg-skyline is not guaranteed to increase, but from Theorem 1 we are sure that the cardinality of general skyline is bigger than skyline. Then we can save the result of kg-skyline and move on to $(k-1)g$-skyline until $\delta$ result is retrieved.

### 4.1   Naive Tuning Algorithm(NTA)

From the above analysis, we know that we can use the technique of k-skyline and general skyline to tune the size of skyline. But to compute $\delta$-skyline, we have to know first the size of skyline and then decide to implement which technique to control the size. The estimation of skyline cardinality is another complex work, so our naive algorithm is to compute skyline first and decide to use which technique. To save the work in computation of skyline we introduce a theorem that will lower the cost of computing $\delta$-skyline.

**Lemma 1.** $k\text{-}skyline \subseteq skyline$.

Proof: if p∈k-skyline and $p \notin$ skyline, then $\exists q \succ p$ and $q \underset{k}{\succ}$ p, so $p \notin$ k-skyline, this contradicts with assumption.   □

**Theorem 3.** *k-skyline can be derived from skyline*

---

**Algorithm 1.** Naive Tuning Algorithm

---

**Input:** data set S on dimension D;
      $\delta$: number of skyline
**Output:** R: $\delta$-skyline;
1: compute skyline on D;
2: $k = |D| - 1$;
3: **if** $|skyline| > \delta$ **then**
4:   **for** $k >= 1$ **do**
5:     compute k-skyline;
6:     **if** $|k - skyline| <= \delta$ **then**
7:       break;
8:     **end if**
9:     k=k-1;
10:     R=k-skyline;
11:   **end for**
12: **else**
13:   **if** $|skyline| < \delta$ **then**
14:     R=$|skyline|$;
15:     **for** $k >= 1$ **do**
16:       **for** any subspace U that $|U| = k$ **do**
17:         compute skyline on U;
18:         R=R + skyline on U;
19:         **if** $|R| >= \delta$ **then**
20:           return R;
21:         **end if**
22:       **end for**
23:     **end for**
24:   **else**
25:     R=skyline;
26:   **end if**
27: **end if**
28: **return** R;

---

Proof: Given p $\in$ k-skyline, if $\exists q \notin$ skyline, so that $q \underset{k}{\succ} p$, then $\exists p' \in$ skyline and $p' \succ q$, so $p' \underset{k}{\succ} p$ and p $\notin$ k-skyline. This contradicts with assumption.     $\square$

From theorem 3, we know that we can compute k-skyline on the result of skyline. And in the proof we show that there is no false positive in the result even if there exists cyclic dominance between points. Our first algorithm is to compute skyline first using SFS and skyline and candidates are stored as an temporary result. Candidate points are those that cannot be strictly dominated, which are possible to be general skyline. If the skyline cardinality is bigger than $\delta$, the algorithm will move on to compute k-skyline on the result of skyline. Or if the skyline cardinality is smaller than $\delta$, the algorithm will move on to find general skyline in the candidates. The detailed algorithm are in Algorithm 1.

## 4.2    Navigating on Skycube(NoS)

Skycube is composed of skyline on different subspace and is quite useful when answering query of users' on different subspace. An example of the skycube lattice on dimension D=A,B,C,D is Figure 4. From the definition of general skyline, it is not hard to infer that kg-skyline are the union of subspace skyline on level k. In addition, we have proved in another work [20] that k-skyline equal to the intersection of subspace skyline on level k. Based on this, we propose our second algorithm based on skycube to compute $\delta$-skyline. The second algorithm showed in algorithm 2 is a top down way navigating on skycube to compute k-dominate and general skyline while k is not fixed to retrieve $\delta$ points. There has been several way to compute skycube from the beginning, such as Stellar [8] or subsky [6]. But our algorithm involves with high dimensional data points, and the construction overhead and storage overhead may be huge. So we adopt the technique of compressed skycube in [3], which can concisely represent the complete skycube and reduce duplicating storage between levels. In this algorithm, we assume that the skycube has already been constructed in advance and be processed as an input.

---

**Algorithm 2.** Navigating on Skycube

---

**Input:** data set S on dimension D;
      $\delta$: number of skyline
      Compressed Skycube on (S,D)
**Output:** R: $\delta$-skyline;
    $k = |D| - 1$;
 2: **if** $|skyline| > \delta$ **then**
      **for** $k >= 1$ **do**
 4:     R=intersection of node in level k;
        **if** $|R| <= \delta$ **then**
 6:        break;
        **end if**
 8:     k=k-1;
      **end for**
10: **else**
      **if** $|skyline| < \delta$ **then**
12:     R=skyline;
      **for** k>= 1 **do**
14:      R=R + union of node on level k;
        **if** $|R| >= \delta$ **then**
16:        break;
        **end if**
18:     k=k-1;
      **end for**
20:   **end if**
    **end if**
22: **return** R;

---

## 5    Performance Evaluation

This section reports our experimental results to validate the effectiveness and efficiency of our proposed algorithms NTA and NoS. First, to validate effectiveness using real-life data, Section 5.1 reports our evaluations over real NBA player statistics. Second, to validate efficiency in extensive problem settings, Section 5.2 reports our evaluations over synthetic datasets of various problem settings. Our experiments were carried out on an Intel 2.4 GHz PC with 512M RAM running Windows XP. Source programs are coded in C++ language.

### 5.1    Experiments on Real NBA Data Set

In this section, we validate the effectiveness of our proposed two algorithms by observing the result of the skyline objects retrieved from the real-life data. In particular, we use the NBA dataset (available from databasebasketball.com) which contains 3570 players with 17 numeric attributes including game points, number of rebounds, assists, steals, blocks and so on from 1999 to 2006. Among these 3570 players, 378 of players are in skyline on $|D|$=17. By running our NTA algorithm, We first compute $\delta$-skyline with $\delta$ equals to 10 on full space. In the real experiments, 14 player are returned and the first ten are showed in table 1. Then we test our algorithm in the condition that the original skyline result is less than $\delta$. We limit the dimension to 2 which contains points and assists of player. This query would be useful if a coach is trying to find a player who has the strongest ability to gain points. The original skyline query will return the first four player in the second column of table-1 and the following 5 player are the result of $\delta$-skyline with $\delta$ equals to 10. From the result on real NBA data, we can conclude that our method can tuning the cardinality of skyline result according to the requirement of user. But a weakness of the method is that it may not be able to tune the result of skyline to exact $\delta$ number(14 and 9 in the example), this is because we still use multi-criterion, the essence of skyline, to determine whether a point is good, not by assigned weights to rank.

**Table 1.** Evaluation on real NBA Data Set, $\delta$=10,$S$=3570

| |D|=17 | |D|=2 |
|---|---|
| KevinGarnett 2003 | Kobe Brant 2005 |
| TimDuncan 2000 | LeBron James 2005 |
| VinceCarter  1999 | AllenIverson 2005 |
| AntoineWalker  1999 | AllenIverson 2004 |
| JermaineO'Neal 2002 | GaryPaton 1999 |
| ShawnKemp 1999 | GaryPaton 2001 |
| ZydrunasIlgauskas 2004 | SteveNash 2005 |
| AmareStoudemire 2002 | SteveNash 2006 |
| RaefLafrentz 2001 | GaryPaton 2000 |
| JoshSmith 2005 | |

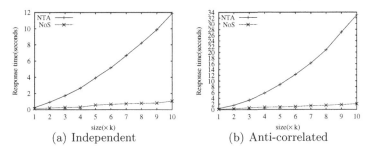

(a) Independent                    (b) Anti-correlated

**Fig. 5.** Performance on varying data size

## 5.2 Experiments on Synthetic Data Set

In this section, we test our algorithms on synthetic data set using the data generator provided by Borzsonyi. We generated two types of data: Independent and anti-correlated data sets with different sizes(from 1k tuples to 50K tuples) and with dimension size up to 15. For the details of data distribution, please refer to [1]. In the following three Figure, we will test the performance of our algorithms when data size, number of dimension and $\delta$ changes. In Figure 5, we test the performance of our algorithms on varying data size, and the dimension and $\delta$ are set to default as 10 and 20. we can see from the figure that the response time of NTA increase sharply as the data size increase. This because that NTA has to compute skyline first and then to compute k-skyline. Although the cost of computing k-skyline on skyline can be ignore, but the time to compute skyline increase sharp as the data size increases. While for the algorithm NoS, the compressed skycube(CSC) are taken as an input, it only need to compute intersection or union of nodes on different level, which is quite fast. So its performance increase slightly when the points in the union of nodes increase.

In Figure 6, we study the effect of dimensionality. The dimension are varying from 2 to 15, and cardinality of data size is 5K and $\delta$ equals to 20. We can see from the figure that the performance of NTA increase first and then drops at dimension of 4. This is because that on dimension 2 and 3,the skyline size is

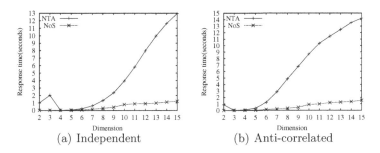

(a) Independent                    (b) Anti-correlated

**Fig. 6.** Performance on varying dimension

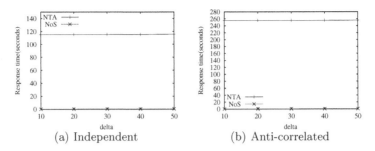

**Fig. 7.** Performance on varying $\delta$

smaller than $\delta$, so NTA runs to find skyline on subspaces which cost more, but after dimension is bigger than 4, the size of skyline is more than $\delta$ then NTA starts to compute k-skyline to decrease the cardinality which is ignorable as we mentioned above.

In figure 7, we evaluate the effect of $\delta$. This time we enlarge the data size to 100K and dimension cardinality to 10 to test the performance of NTA and NoS when the cardinality of skyline is huge. We can see from the figure that although the time increase to compute skyline on different distribution of data( independent or anti-correlated), the response time is nearly constant when the $\delta$ varies. The reason for NoS to be costant is the same as we explained above. For NTA, it is because in the implementation of NTA, we use some auxiliary data structure to store the max k of a skyline that can be k-dominated by other skylines. Moreover it is almost instant after one iteration of computing max k for each skyline and to determine a proper k to limit the size of $\delta$-skyline equal to or smaller than $\delta$.

## 6    Conclusions

In this paper, we define the concept of general skyline, which include all the subspace skylines. We show that general skyline is a meaningful extending of skyline operator when cardinality of skyline operator is not sufficient. And we are the first to propose a method to tune the cardinality of skyline operator in both directions, to either increase or decrease according to the requirement of the user, without any user interference such as assigned weights or scoring functions. By our method, the output size of skyline operator can be controlled and no ranking functions need to be specified by users. In addition, the result is independent of the scales of different dimensions. In short, the advantage of skyline over top-k query can be remained to achieve our goal. Extensive experiments on both the synthetic and real NBA data have demonstrated the efficiency and effectiveness of our proposed algorithms. In the future, we will do more work on improving our algorithms to compute $\delta$-skyline more efficiently.

# References

1. Borzonyi, S., Kossmann, D., Stocker, K.: The skyline operator. In: Proc. of ICDE, pp. 421–430 (2001)
2. Tian, X., Zhang, D., Tao, Y.: On Skyling with Flexible Dominance Relation. In: Proc. of ICDE (2008)
3. Xia, T., Zhang, D.: Refreshing the Sky: The Compressed Skycube with Efficient Support for Frequent Updates. In: Proc. of SIGMOD (2006)
4. Lin, X., Yuan, Y., et al.: Selecting Stars: The k Most Representative Skyline Operator. In: Proc. of ICDE (2007)
5. Yiu, M., Mamoulis, N.: Efficient Processing of Top-k Dominating Queries on Multi-Dimensional Data. In: Proc. of VLDB (2007)
6. Tao, Y., Xiao, X., Pei, J.: Efficient Skyline and Top-k Retrieval in Subspaces. In: Proc. of TKDE, pp. 1072–1088 (2007)
7. Chan, Y., Jadadish, H., et al.: Finding k-Dominant Skylines in High Dimensional Space. In: Proc. of SIGMOD, pp. 503–514 (2006)
8. Pei, J., Fu, A.W., et al.: Computing Compressed Multidimensional Skyline Cubes Efficiently. In: Proc. of ICDE (2007)
9. Yuan, Y., Lin, X., et al.: Efficient Computation of the Skyline Cube. In: Proc. of VLDB (2005)
10. Papadias, D., Tao, Y., et al.: Progressive Skyline Computation in Database Systems. In: ACM Transactions on Database Systems, pp. 41–82 (2005)
11. Pei, J., Jin, W., et al.: Catching the Best Views of Skyline: A Semantic Approach Based on Decisive Subspaces. In: Proc. of VLDB (2005)
12. Chan, C.Y., Jagadish, H., et al.: On High dimensional Skylines. In: Ioannidis, Y., Scholl, M.H., Schmidt, J.W., Matthes, F., Hatzopoulos, M., Böhm, K., Kemper, A., Grust, T., Böhm, C. (eds.) EDBT 2006. LNCS, vol. 3896. Springer, Heidelberg (2006)
13. Balke, W.-T., Guntzer, U., Zheng, J.X.: Efficient distributed skylining for web information systems. In: Bertino, E., Christodoulakis, S., Plexousakis, D., Christophides, V., Koubarakis, M., Böhm, K., Ferrari, E. (eds.) EDBT 2004. LNCS, vol. 2992, pp. 256–273. Springer, Heidelberg (2004)
14. Chomicki, J., Godfrey, P., Gryz, J., Liang, D.: Skyline with presorting. In: Proc. of ICDE, pp. 717–719 (2003)
15. Kossmann, D., Ramsak, F., Rost, S.: Shooting Stars in the Sky: An Online Algorithm for Skyline Queries. In: Proc. of VLDB, pp. 275–286 (2002)
16. Lo, E., Ip, K., Lin, K.-I., Cheung, D.: Progressive Skylining over Web-Accessible Database. DKE 57(2), 122–147 (2006)
17. Jin, W., Han, J., Ester, M.: Mining Thick Skyline Over Large Databases. In: Proc. of PKDD (2004)
18. Lin, X., Yuan, Y., Wang, W., Lu, H.: Stabbing the Sky: Efficient Skyline Computation over Sliding Windows. In: Proc. of ICDE, pp. 502–513 (2005)
19. Papadias, D., Tao, Y., Fu, G., et.,, al,: An Optimal and Progressive Algorithm for Skyline Querie. In: Proc. of SIGMOD, pp. 467–478 (2003)
20. Xin, J., Wang, G., et al.: Energy Efficient computing of k-dominant skline on Sensor Network. In: Proc. of NDBC (2008)
21. Tan, K.-L., Eng, P.-K., Ooi, B.C.: Efficient progressive skyline computation. In: Proc. Of VLDB, pp. 301–310 (2001)
22. Tao, Y., Papadias, D.: Maintaining Sliding Window Skylines on Data Streams. TKDE 18(3), 377–391 (2006)
23. Wu, M., Xu, J., Tang, X., Lee, W.-C.: Monitoring Top-k query in wireless sensor network. In: Proc. of ICDE, p. 143 (2006)

# An HMM Approach to Anonymity Analysis of Continuous Mixes

Zhen Ling, Junzhou Luo, and Ming Yang

School of Computer Science and Engineering, Southeast University, Nanjing, P.R. China
{zhen_ling,jluo,yangming2002}@seu.edu.cn

**Abstract.** With the increasing requirement of privacy protection, various anonymity communication systems are designed and implemented. However, in the current communication infrastructure, traffic data can be gathered at moderate cost by adversary. Based on the traffic data, they can easily correlate the input links with output links by applying powerful traffic analysis techniques. In this paper, a Hidden Markov Model (HMM) approach is proposed to analyze one of the important anonymity systems, continuous mixes, which individually delays messages instead of processing batch messages. This approach consists of two parts, arrival traffic model and departure traffic model based on HMM, which capture the mean rates of the arrival and departure messages respectively. By using this approach to analyze anonymity of continuous mixes, a successful anonymity analysis can not be guaranteed, especially while the arrival traffic rate is greater than the departure traffic rate. In order to achieve better anonymity results, a new countermeasure is proposed, which inserts a minimum number of dummy traffic flows to ensure better anonymity of continuous mixes and protects users against various traffic analyses.

**Keywords:** continuous mixes, traffic analysis, Hidden Markov Model (HMM), anonymity, dummy traffic.

## 1 Introduction

It was not before year 2000 that anonymity started to receive enough attention from research communities, although anonymity research started in the early 1980's with David Chaum's paper on untraceable email [1]. Mixes which are applied in many anonymous communication systems, such as timed, threshold and pool mixes, have been extensively researched [2, 3]. Whereas the continuous mixes, proposed by Kesdogan et al [4], individually delay messages instead of processing batch input messages in rounds, the analyzing approach of this type of mixes is not similar to previous mix-based anonymity systems.

Many types of passive attacks on anonymity systems have been proposed, and they are mainly based on traffic analysis techniques and hard to detect. Timing analysis attack [5, 6], the adversary can use information on the delay by the mixes in every link to correlate input link with output one. Intersection attack [7] consists of intersecting anonymity sets of consecutive messages sent or received by a user. Disclosure

Y. Ishikawa et al. (Eds.): APWeb 2008 Workshops, LNCS 4977, pp. 232–243, 2008.
© Springer-Verlag Berlin Heidelberg 2008

attack [8] is a variant of the intersection attack which improves the efficiency by requiring less effort from the attacker. Recently, some more powerful active attacks are presented. Flow marking technique [9] is applied by utilizing a Pseudo-Noise code. And the secret Pseudo-Noise code is difficult for others to detect, therefore the traceback is effectively invisible.

The inventors [4] of continuous propose an analysis of its anonymity, but they have not mentioned the limitation. George Danezis [10] proposes to apply the information-theoretic anonymity metrics to continuous mixes and finds the flaws of continuous mixes. His work cannot dynamically find out the bad anonymity state in the real time. Claudia Diaz et al [11] also identify flaws of continuous mixes, but they have not come up with any efficient countermeasures to provide well anonymity. Andrei Serjantov [12] model continuous mixes as an *M/M/n* process by using standard queuing theory techniques, but the author did not pointed out the limitation of it.

In this paper, we propose a new Hidden Markov Model (HMM) approach to anonymity analysis of continuous mixes. The HMM approach consists of arrival traffic model and departure traffic model. These two models based on HMM can capture the arrival state and departure state respectively. With the help of these two models, we can find out the bad anonymity state in real time. We also present an efficient countermeasure by inserting appropriate dummy traffic to improve anonymity of continuous mixes.

## 2   Anonymity for Continuous Mixes

The mix, designed by Chaum [1] in 1981, takes a number of input messages, and outputs messages in such a way that it is hard to link an output to the corresponding input and the flow of messages. Firstly, the bitwise unlinkability of inputs and outputs can be achieved by using encryption and padding messages to transform the pattern of the traffic. Secondly, the mix changes the timing correlations between the inputs and outputs by delaying and reordering messages in order to make it difficult for an attacker to find a link between an input and an output. A continuous mix achieves this by delaying each message individually, regardless of the current load.

### 2.1   A Mixing Strategy for Continuous Mixes

The Stop-and-Go mixes are also called continuous mixes, proposed by Kesdogan et al. in [4]. The continuous mixes delay each arrival message according to a random variable that follows the exponential distribution with parameter $\mu$, and then forwards messages individually. Therefore the messages are reordered by the randomness of the delay distribution, and the delay characteristic of the continuous mixes is:

$$f(t) = \mu e^{-\mu t}. \tag{1}$$

This delay probability density function $f(t)$ represents the probability a message leaves the continuous mixes in a time interval $t$.

## 2.2 Anonymity Metrics

Similar to [10], to facilitate our analysis, we assume that the rate of messages to the mix is in Poisson distribution with rate $\lambda$. To measure continuous mixes's sender anonymity, we will adopt the metric introduced in [13], which is based on defining a random variable that describes the possible senders of a message and calculating the entropy of its underlying probability distribution. The entropy is a measure of the anonymity provided, and can be represented as the amount of information an attacker is missing to deterministically link the messages to a sender. The sender anonymity can be calculated by Equation (2):

$$H = -\log \frac{\lambda e}{\mu}. \tag{2}$$

When a departure rate $\mu$ of a message queue is greater than the arrival rate $\lambda$, the continuous mixes would not provide much anonymity most of the time. The average time while a message would stay in the mix is $1/\mu$, while the average time between message arrivals is $1/\lambda$, which is greater than $1/\mu$. Therefore the mix would behave on average as a first-in first-out queue, and then the continuous mixes can rarely ensure the anonymity.

## 2.3 Threat Model

In this paper, we assume that the adversary empolys a classical timing analysis attack [5, 6]. The mix is observed by the adversary, who can observe input and output links of the whole mix network. Since traffic is not altered, this is a type of passive attack. The attacker knows the mix's infrastructure and all internal parameters of the mix.

Because message packets are respectively encrypted and padded or split to the same size, the attacker cannot find any correlation based on packet content or packet size on an input link with another packet on the output link.

To simplify the following discussion, we assume that dummy traffic is not used in continuous mix network.

# 3   Arrival Traffic Model Based on HMM

## 3.1   Observation Sequence of Arrival Traffic

We split the whole time into a lot of continuous time unit $T$, in a time unit $T$, $K$ messages arrive at the mix, where $K$ obeys a Poisson distribution with parameter $\lambda$. Each message arrives at time unit T obeys a uniform distribution. We make $\{o_t = K\}$ ($t \geq 1$ and $K \in Z$) as an observation sequence of arrival traffic to the mix.

## 3.2   The Model

According to the research of Internet traffic [14, 15], we can consider that the arrival traffic to the mix can be well modeled by an HMM with appropriate hidden variables that capture the current state of the arrival rate. Therefore we propose arrival traffic model based on HMM.

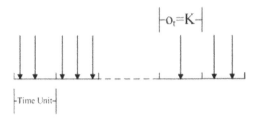

**Fig. 1.** An observation sequence of arrival traffic

We assume there are $N$ discrete states of the arrival rate, and $q_t$ denotes the state of arrival traffic at time slot $t$, with $q_t \in \{s_1, s_2, ..., s_N\}$, where $s_i$ is the $i$th state, among $N$ possible ones. The set of parameters characterizing the model is $\Omega = \{A, B, \pi\}$.

$A$ is the state transition matrix, where $A = \{a_{ij}\} = P(q_{t+1} = s_j | q_t = s_i)$. It represents the probability of the state transmit from $s_i$ to $s_j$. $o_t$ is a discrete random variable whose dynamic behavior is governed by the transition matrix A;

$B$ is the state-conditioned the number of arrival messages probability, where $B = b(k|q_i) = b(o_t = k|q_t = s_i)$. It represents the probability of observing the number of arrival messages at time step $t$ in the state $s_j$. $o_n$ is also a discrete random variable that, given $\{q_t = s_i\}$, is characterized by Poisson distribution function:

$$b\left(o_t = k \mid q_t = s_i\right) = \frac{\lambda_i^k}{k!} e^{-\lambda_i}. \tag{3}$$

$\pi$ is the initial state distribution, where $\pi = \{\pi_1, \pi_2, ..., \pi_N\}$, $\pi_i = P(q_1 = s_i)$ and $\sum_{i=1}^{N} \pi_i = 1$.

### 3.3 Learning the Model Parameters

We adopt EM algorithm [16] for learning the model parameter from the observation sequence. Furthermore, such a model would be useful only if the rate of arrival traffic holds on the same stochastic behavior observed during the training step, i.e., the rate of arrival traffic alternates different short-term behavior while showing a long-term stationary.

The EM algorithm is an optimization procedure that allows learning of a new set of parameters for a stochastic model according to improvements of the likelihood of a given sequence of observable variables. For HMM this optimization technique reduces to the Forward-Backward algorithm [17, 18].

Specifically, given a sequence of arrival traffic with observable variables $o = (o_1, o_2, ..., o_T)$, referred to as the training sequence, we search the set of parameters $\Omega_{max}$ such that the likelihood of the training sequence is maximum:

$$\begin{cases} L(o;\Omega) = P(o \mid \Omega) \\ \Omega_{max} = \arg\max_{\Omega} \{L(o;\Omega)\} \end{cases}. \tag{4}$$

In order to search a local maximum of the likelihood, we apply the Forward-Backward algorithm with our deduced iterative formula $I(o;\Omega)$ so that $L(o;\hat{\Omega}) \geq L(o;\Omega)$. Furthermore, we can find the global solution by repeating starts with different initial conditions.

According to (3), the arrival traffic is characterized by Poisson distribution function. Therefore, $\Omega = \{A, B, \pi\}$ is equal to $\Omega = \{A, \lambda, \pi\}$, where $\lambda = \{\lambda_1, \lambda_2, ..., \lambda_N\}$ denotes state-conditioned arrival rate means, i.e. $\lambda_i = E(o_t \mid \{q_t = s_i\})$.

The algorithm works as follows:

(1)   Set input parameters: This includes the number of states $N$, the training sequence $o = (o_1, o_2, ..., o_T)$, and the stopping parameter $\varepsilon$.

(2)   Initialization step: the set of parameter $\Omega = \{A, \lambda, \pi\}$ is initialized randomly.

(3)   Iterative step:

a) Calculate the new set of parameter $\hat{\Omega} = I(o;\Omega)$, and then calculate the likelihood $L(o;\hat{\Omega})$.

b) if $L(o;\hat{\Omega}) - L(o;\Omega) \leq \varepsilon$, go to (4), else go to (3a).

(4)   The algorithm stop.

The Forward-Backward Procedure [17, 18] : Consider the forward variable $\alpha_t(j)$ and backward variable $\beta_t(i)$ respectively defined as

$$\alpha_t(j) = P(o_1 o_2 \cdots o_t, q_t = s_i \mid \Omega) = \left[ \sum_{i=1}^{N} \alpha_{t-1}(i) a_{ij} \right] b_j(o_t). \tag{5}$$

$$\beta_t(i) = P(o_{t+1} o_{t+2} \cdots o_T \mid q_t = s_i, \Omega) = \sum_{j=1}^{N} a_{ij} b_j(o_{t+1}) \beta_{t+1}(j). \tag{6}$$

According to formulas (5) and (6), we can give reestimation formulas [19] of the set of parameters $\hat{\Omega} = \{\hat{A}, \hat{\lambda}, \hat{\pi}\}$ as follows:

$$\hat{a}_{ij} = \frac{\sum\limits_{t=1}^{T-1} \xi_t(i,j)}{\sum\limits_{t=1}^{T-1} \gamma_t(i)} = \frac{\sum\limits_{t=1}^{T-1} \alpha_t(i) a_{ij} b_j(o_{t+1}) \beta_{t+1}(j)}{\sum\limits_{t=1}^{T-1} \alpha_t(i) \beta_t(i)}. \tag{7}$$

$$\hat{\lambda}_i = \frac{\sum\limits_{t=1}^{T} \gamma_t(i) \cdot o_t}{\sum\limits_{t=1}^{T} \gamma_t(i)} = \frac{\sum\limits_{t=1}^{T} \alpha_t(i) \beta_t(i) \cdot o_t}{\sum\limits_{t=1}^{T} \alpha_t(i) \beta_t(i)}. \tag{8}$$

$$\hat{\pi}_i = \frac{\sum_{t=1}^{T} \gamma_t(i)}{\sum_{i=1}^{N}\sum_{t=1}^{T} \gamma_t(i)} = \frac{\sum_{t=1}^{T} \alpha_t(i)\beta_t(i)}{\sum_{i=1}^{N}\sum_{t=1}^{T} \alpha_t(i)\beta_t(i)}. \tag{9}$$

## 4  Departure Traffic Model Based on HMM

### 4.1  Observation Sequence of Departure Traffic

We assume that the messages from arrival mix are numbered as $v$ ($v$ = 1, 2, …); $\alpha_v$ and $\beta_v$ are the arrival time and the departure time of the $v$th message. As the messages depart the mix, the delay of the $v$th messages in mix is $\sigma_v = \beta_v - \alpha_v$, which is distributed according to an exponential distribution with parameter $\mu$. We make $\{x_v = \sigma\}$ ($v \geq 1$ and $\sigma > 0$) as an observation sequence of departure traffic.

### 4.2  The Model

Similar to arrival traffic model, we assume there are $M$ discrete states of the departure rate, and $y_v$ denotes the state of the departure at time step $v$, with $y_v \in \{r_1, r_2, …, r_M\}$, where $r_i$ is the $i$th state, among $M$ possible ones. The set of parameters characterizing the model is $\Psi = \{A, \pi, f_1(\sigma), f_2(\sigma), …, f_M(\sigma)\}$.

$A$ is the state transition matrix, where $A = \{a_{ij}\} = P(y_{v+1} = r_j | y_v = r_i)$. It represents the probability of the state transmit from $r_i$ to $r_j$. $y_v$ is a discrete random variable whose dynamic behavior is governed by the transition matrix $A$;

$f_i(\sigma)$ is the state-conditioned delay probability density function, where $P(\sigma_v > t \mid y_v = r_i) = \int_{t}^{+\infty} f_i(\sigma) d\sigma$. It represents the probability of observing the delay of messages at time step $v$ in the state $r_j$. $y_v$ is a continuous random variable that, given $\{y_v = r_i\}$, is characterized by exponential probability density function:

$$f_i(t) = \mu e^{-\mu t}. \tag{10}$$

$\pi$ is the initial state distribution, where $\pi = \{\pi_1, \pi_2, …, \pi_M\}$, $\pi_i = P(y_1 = r_i)$ and $\sum_{i=1}^{M} \pi_i = 1$.

### 4.3  Learning the Model Parameters

The delay dynamics of the messages in the mix suggests the introduction of a hidden-state variable carrying information about departure situation of the messages. The state variable stochastically influences delay of the messages. Since our knowledge about the state can only be inferred from observation of delays, and there is no way to access it directly. We will apply later EM algorithm to capture the hidden states in Section 6.

According to (10), the departure traffic is characterized by exponential distribution function. Therefore, $\Psi = \{A, \pi, f_1(\sigma), f_2(\sigma), ..., f_M(\sigma)\}$ is equal to $\Psi = \{A, \pi, \mu\}$, where $\mu = \{\mu_1, \mu_2, ..., \mu_M\}$ denotes state-conditioned departure rate means, i.e., $\mu_i = E(x_v | \{y_v = r_i\})$.

With the parameter of the number of hidden-state $M$, the training sequence $x = (x_1, x_2, ... , x_V)$ and $\Psi = \{A, \pi, \mu\}$, we can deduce the forward variable $\alpha_v'(j)$ and backward variable $\beta_v'(i)$ respectively as follows:

$$\alpha_v(j) = P(x_1 x_2 \cdots x_v, y_v = r_i | \Psi) = \left[ \sum_{i=1}^{M} \alpha_{v-1}(i) a_{ij} \right] b_j(x_v). \tag{11}$$

$$\beta_v(i) = P(x_{v+1} x_{v+2} \cdots x_V | y_v = r_i, \Psi) = \sum_{j=1}^{M} a_{ij} b_j(x_{v+1}) \beta_{v+1}(j). \tag{12}$$

According to formulas (5) and (6), we can give reestimation formulas of the set of parameters $\hat{\Psi} = \{\hat{A}, \hat{\pi}, \hat{\mu}\}$ as follows:

$$\hat{a}_{ij} = \frac{\sum_{v=1}^{V-1} \xi_v(i,j)}{\sum_{v=1}^{V-1} \gamma_v(i)} = \frac{\sum_{v=1}^{V-1} \alpha_v(i) a_{ij} b_j(x_{v+1}) \beta_{v+1}(j)}{\sum_{v=1}^{V-1} \alpha_v(i) \beta_v(i)}. \tag{13}$$

$$\hat{\pi}_i = \frac{\sum_{v=1}^{V} \gamma_v(i)}{\sum_{i=1}^{M} \sum_{v=1}^{V} \gamma_v(i)} = \frac{\sum_{t=1}^{V} \alpha_v(i) \beta_v(i)}{\sum_{i=1}^{M} \sum_{v=1}^{V} \alpha_v(i) \beta_v(i)}. \tag{14}$$

$$\hat{\mu}_i = \frac{\sum_{v=1}^{V} \gamma_v(i) \cdot x_v}{\sum_{v=1}^{V} \gamma_v(i)} = \frac{\sum_{v=1}^{V} \alpha_v(i) \beta_v(i) x_v}{\sum_{v=1}^{V} \alpha_v(i) \beta_v(i)}. \tag{15}$$

The procedure of EM algorithm is similar with section 4.3, we will not present it again. To implement this algorithm, it requests proper scaling to avoid underflow [18].

## 5   Empirical Evaluation

In this section, we evaluate the effectiveness of the continuous mix with our proposed traffic analysis approach based on HMM. We see the failure of anonymity when a message queue with a departure rate $\mu$ is larger than the arrival rate $\lambda$, the continuous mix would not provide much anonymity most of the time. Measures of delays have been performed on our experimental continuous mix.

### 5.1 Departure Traffic Analysis Based on HMM

We set the number of the hidden-state is 3, and it represents that the departure rate are low, normal and high respectively: $\mu_1 \le \mu_2 \le \mu_3$. A training sequence of $V$=1000 samples is applied, and the EM algorithm convergence is reached after a few iterations. Fig.2 shows a typical trend of log-likelihood evolution during the EM learning procedure. It shows that 7 iterations

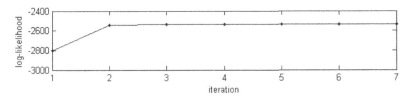

**Fig. 2.** Log-likelihood trend in the EM algorithm

Fig.3 shows a portion of delay sequence. To simplify our analysis, we take 100 samples as an example.

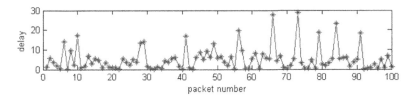

**Fig. 3.** Delays of the messages in continuous mix

In order to verify how the hidden-state variable $y_v$ capture the current departure traffic state of continuous mix, a Viterbi algorithm [18] is applied to the training sequence, which searches the most likely state sequence $x = (x_1, x_2, \dots, x_V)$. Fig.4 shows the state sequence obtained by applying of the Viterbi algorithm on the three state trained models. The trained model well describes the departure traffic state of continuous mix. And we get the average departure rate $(\mu_1, \mu_2, \mu_3) = (1.9695, 5.1211, 7.6998)$ and the steady-state probability $(\pi_1, \pi_2, \pi_3) = (0.1183, 0.5288, 0.3528)$. From these results, we know that most of the messages stay in state 2.

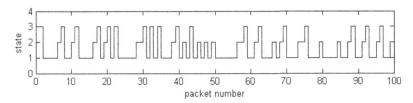

**Fig. 4.** State of departure traffic in continuous mix

## 5.2 Arrival Traffic Analysis Based on HMM

Similar with departure traffic analysis, we also set the number of the hidden-state is 3, and it represents that the arrival rate are low, normal and high respectively: $\lambda_1 \leq \lambda_2 \leq \lambda_3$. A training sequence of $T=1000$ samples is applied.

Fig.5 shows a portion of arrival messages sequence in continuous mix.

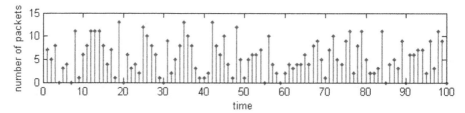

**Fig. 5.** Arrival traffic in continuous mix

With a Viterbi algorithm [18] applied to the training sequence, the most likely state sequence $o = (o_1, o_2, \dots , o_T)$ is captured. Fig.6 shows the state sequence obtained by applying of the Viterbi algorithm on the three state trained models. Furthermore, we get the average arrival rate $(\lambda_1, \lambda_2, \lambda_3) = (3.4285, 6.7431, 8.2091)$ and the steady-state probability $(\pi_1, \pi_2, \pi_3) = (0.2397, 0.5163, 0.2439)$. From these results, we know that most of the arrival traffic stays in state 2.

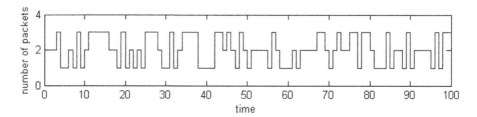

**Fig. 6.** State of arrival traffic in continuous mix

## 5.3 Anonymity Analysis of Continuous Mixes

Followed by the results from the experiment, we can get a sequence $\mu_1 < \lambda_1 < \mu_2 < \lambda_2 < \mu_3 < \lambda_3$. While a message queue with a departure rate $\mu$ is larger than the arrival rate $\lambda$, the mix would behave on average as a first-in first-out queue. In this case, the attacker can easily link between inputs and outputs. Therefore, the continuous mix would not provide much anonymity most of the time.

With the permutation and combination, we can obtain 9 anonymity levels, which the entropy $H_{best} = -\log \dfrac{\lambda_3 e}{\mu_1}$ achieves the best anonymity, i.e., the arrival traffic is in state 3, whereas the departure traffic is in state 1.

# 6  Dummy Traffic Policy

After the anonymity analysis of continuous mix in section 5, we find out the limitation of continuous mix. In this section, we adopt a relatively efficient and effective countermeasure.

Since the rates of traffic among the inputs and outputs of the continuous mix are different, we appropriately insert dummy traffic to make the anonymity achieve the best. Because of the heavy consumption of bandwidth of dummy traffic, the challenge is to insert a minimum number of dummy traffic.

Our countermeasure algorithm is illustrated as follows:

```
While (1) do
    if (λ = λ₃) & (μ = μ₁)
        relay the packets;
    else
        send dummy packets w = (λ₁/μ₃)μT - λT ;
end
```

We make the best anonymity as the standard and assume that the number of dummy packets inserted into mix is $w$. While the arrival state is not state 3 and the departure state is not state 1, we insert dummy traffic into mix. By inserting dummy packets $w$, the anonymity can be calculated as follows:

$$H = -\log\frac{(w/T + \lambda)e}{\mu} . \tag{16}$$

We order $H_{best} = H$, then $w$ equals to $\dfrac{\lambda_1}{\mu_3}\mu T - \lambda T$ . It is efficient because packets are forwarded based on both the arrival rate and departure rate. We keep the continuous mix in the best anonymity level by inserting minimums dummy traffic.

# 7  Conclusion

In this paper, an HMM approach is introduced for anonymity analysis of continuous mixes. The arrival traffic mode and departure traffic model based on HMM are proposed to capture the arrival and departure states respectively. We can find out the bad anonymity state of continuous mixes in the real time by our traffic analysis models,. We also propose a new countermeasure by inserting appropriate dummy traffic to improve anonymity. A new countermeasure algorithm is given and we also work out how many dummy packets are needed in different state. Through theoretical and empirical analysis of our HMM approach, we significantly contribute to previous research on anonymity by inserting dummy packets.

Our further research is to further explore more efficient countermeasures and to apply our extended HMM approach to various anonymity systems.

**Acknowledgments.** This work is supported by National Natural Science Foundation of China under Grants No.90604004, Jiangsu Provincial Natural Science Foundation of China under Grants No.BK2007708, Jiangsu Provincial Key Laboratory of Network and Information Security under Grants No.BM2003201, Key Laboratory of Computer Network and Information Integration (Southeast University), Ministry of Education under Grants No.93K-9 and International Science and Technology Cooperation Program of China.

# References

1. Chaum, D.L.: Untraceable electronic mail, return addresses, and digital pseudonyms. Communications of the ACM 24(2), 84–88 (1981)
2. Díaz, C., Serjantov, A.: Generalising mixes. In: Dingledine, R. (ed.) PET 2003. LNCS, vol. 2760. Springer, Heidelberg (2003)
3. Serjantov, A., Dingledine, R., Syverson, P.F.: From a Trickle to a Flood: Active Attacks on Several Mix Types. In: Petitcolas, F.A.P. (ed.) IH 2002. LNCS, vol. 2578, pp. 36–52. Springer, Heidelberg (2003)
4. Kesdogan, D., Egner, J., Büschkes, R.: Stop-and-Go-MIXes Providing Probabilistic Anonymity in an Open System. In: Aucsmith, D. (ed.) IH 1998. LNCS, vol. 1525, Springer, Heidelberg (1998)
5. Levine, B.N., Reiter, M.K., Wang, C., et al.: Timing Attacks in Low-Latency Mix Systems. In: Juels, A. (ed.) FC 2004. LNCS, vol. 3110. Springer, Heidelberg (2004)
6. Shmatikov, V., Wang, M.H.: Timing Analysis in Low-Latency Mix Networks: Attacks and Defenses. In: Gollmann, D., Meier, J., Sabelfeld, A. (eds.) ESORICS 2006. LNCS, vol. 4189, pp. 18–33. Springer, Heidelberg (2006)
7. Agrawal, D., Kesdogan, D., Penz, S.: Probabilistic treatment of MIXes to hamper traffic analysis. In: Proceedings of Symposium on Security and Privacy 2003, pp. 16–27 (2003)
8. Danezis, G., Diaz, C., Troncoso, C.: Two-sided Statistical Disclosure Attack. Privacy Enhancing Technologies, 30–44 (2007)
9. Yu, W., Fu, X., Graham, S., et al.: DSSS-Based Flow Marking Technique for Invisible Traceback. In: Proceedings of the 2007 IEEE Symposium on Security and Privacy, pp. 18–32 (2007)
10. Danezis, G.: The Traffic Analysis of Continuous-Time Mixes. Privacy Enhancing Technologies, 35–50 (2005)
11. Díaz, C., Sassaman, L., Dewitte, E.: Comparison Between Two Practical Mix Designs. In: Samarati, P., Ryan, P.Y.A., Gollmann, D., Molva, R. (eds.) ESORICS 2004. LNCS, vol. 3193, pp. 141–159. Springer, Heidelberg (2004)
12. Serjantov, A.: A Fresh Look at the Generalised Mix Framework. In: Borisov, N., Golle, P. (eds.) PET 2007. LNCS, vol. 4776, p. 17. Springer, Heidelberg (2007)
13. Serjantov, A., Danezis, G.: Towards an Information Theoretic Metric for Anonymity. Privacy Enhancing Technologies, 259–263 (2003)
14. Salvo Rossi, P., Romano, G., Palmieri, F., Iannello, G.: Joint end-to-end loss-delay hidden Markov model for periodic UDP traffic over the Internet. IEEE Transactions on Signal Processing 54(2), 530–541 (2006)
15. Muscariello, L., Mellia, M., Meo, M., et al.: Markov models of internet traffic and a new hierarchical MMPP model. Computer Communications 28(16), 1835–1851 (2005)

16. Bilmes, J.A.: A Gentle Tutorial of the EM Algorithm and its Application to Parameter Estimation for Gaussian Mixture and Hidden Markov Models. International Computer Science Institute 4 (1998)
17. Liporace, L.: Maximum likelihood estimation for multivariate observations of Markov sources. IEEE Transactions on Information Theory 28(5), 729–734 (1982)
18. Rabiner, L.R.: A tutorial on hidden Markov models and selected applications inspeech recognition[J]. Proceedings of the IEEE 77(2), 257–286 (1989)
19. Leroux, B.G., Puterman, M.L.: Maximum-penalized-likelihood estimation for independent and Markov-dependent mixture models. Biometrics 48(2), 545–558 (1992)

# Author Index

Chai, Yaohui 165
Chen, Dongling 172
Chen, Li 65

Ding, Dabin 220
Ding, Zhiming 1
Domínguez, Eladio 122

Feng, Yucai 85
Fernandes, Paulo 110
Fu, Haiming 182

Gao, Guangxia 165

Hansen, David 74
Hao, Yanan 210
Harkiolakis, Nicholas 54
Hicks, David 54
Huang, Fuchun 99
Huang, Jianmei 220

Jiang, Yi 65
Jin, Dianchuan 210
Jin, Zhen Ai 155

Karunanithi, Mohan 74

Lawley, Michael 74
Lee, Sang-Eon 155
Li, Bixin 144
Li, Jun 182
Li, Lihong 210
Li, Min 192
Li, Peng 134
Liang, Junjie 85
Ling, Zhen 232
Liu, Baoxiang 210
Liu, WeiYi 87
Lopes, Lucelene 110
Luo, Junzhou 232

Maeder, Anthony 74
McBride, Simon 74
Memon, Nasrullah 54
Morgan, Gary 74
Muccini, Henry 144

Nie, Guangli 182

Pang, Chaoyi 74
Peng, Ni 30
Pérez, Beatriz 122

Qureshi, Abdul Rasool 54

Salvado, Olivier 74
Sarela, Antti 74
Scalabrin, Edson Emilio 110
Shi, Yong 165
Sun, Mingjie 144

Tang, Xijin 17, 42
Thongkam, Jaree 99

Wang, Daling 172, 201
Wang, Guoren 220
Wang, Hua 192
Wang, Ling 65
Wang, Shan 134

Xin, Junchang 220
Xu, Guandong 99

Yan, Hongcan 210
Yan, Hongliang 30
Yang, Haizhen 30
Yang, Ming 232
Yoo, Kee Young 155
Yoshida, Taketoshi 42
Yu, Ge 172, 201
Yue, Kun 87

Zapata, María A. 122
Zhang, JiaDong 87
Zhang, Lingling 182
Zhang, Pengcheng 144
Zhang, Wen 42
Zhang, Yanchun 99
Zhang, Yifei 201
Zhang, Zhiwang 165
Zhu, Qing 134